# Across the Wing

## Stories of Navy Carrier Combat Squadrons in the Vietnam Theatre

by

### Dan Heller

Front and back cover photos:
Captain Steve Kuhar, USNR (ret.)

Back cover author photo:
Cameron Gause

*Across the Wing* Paperback
P3
ISBN 979-8-9883383-0-7
Copyright © 2023 Dan Heller All rights reserved.

Dedicated to my Father
Samuel Heller
Tec 5 U.S. Army World War II
The greatest man I will ever know

# Table of Contents

# Introduction

One of my tasks at Lyon Air Museum in Santa Ana, California is interviewing fellow docents on their military service for preservation in the museum archives and Library of Congress. One such interview in the spring of 2018 was with retired United Airlines Captain Larry Liguori, who had been a Navy F-8 Crusader fighter pilot during the Vietnam conflict before a long and distinguished career as a jumbo pilot with UAL. As we chatted about the terrible losses the United States suffered in Southeast Asia he mentioned that one of his friends from high school in Pasadena, California disappeared without a trace into the jungles of North Vietnam.

Navy VA-35 *Black Panthers* A-6 Intruder Bombardier/Navigator Lieutenant James Kelly Patterson had been shot down near Hanoi on 19 May 1967 after flying off the aircraft carrier *Enterprise*. The pilot of the aircraft he believed had been killed, while Kelly had managed to eject and contact rescue forces before disappearing. I jotted down the name on my notepad and continued the interview.

Later, I casually queried the internet on James Kelly Patterson. Expecting a search result of basic information, what came back was an incredible story of courage, heroism, and tragedy. I also discovered that the pilot of the Intruder, Lieutenant Commander Eugene "Red" McDaniel, had survived as a POW, and was one of the most revered, honored, and respected alumni of the "Hanoi Hilton".

I found many personal and heartfelt tributes to Kelly from those who knew him and missed him terribly. While most were from Naval Academy classmates and others he had served with, one missive that stood out was from his younger brother George, known to family and friends as "Luck". In his heartfelt tribute Luck disclosed that at the time his older brother was shot down, he had been commanding a Marine Corps rifle platoon in South Vietnam and Kelly had paid him a surprise visit in Hoi An about a month before he was lost. It was the last time the two brothers saw each other.

Intrigued and curious, I located an e-mail address for Luck and reached out to him. Within a few hours, he replied that not only did he remember Larry, but as it turns out he had taken his younger sister out on a date in high school. In a fortunate twist of serendipity, Luck lived only about ten miles away from me. We made arrangements to meet at the museum where Larry and I would give him a tour, followed by lunch. The two men had not seen each other for over fifty years, and my excitement over meeting Luck and reuniting the two old acquaintances was palpable. Both men had a great time reminiscing, talking about growing up in Pasadena and what had become of their lives in the decades since. As we said our goodbyes Luck called me over to his car and handed me a thumb drive. "This is Kelly's story," he said as he handed me the device, "more than fifty years of effort. Please be careful with it."

In reviewing the contents of the thumb drive I was astonished to discover that from the time Kelly was shot down to the present day Luck has been on an unrelenting quest to find out the truth of what happened to his brother. From doggedly pursuing government investigators to performing painstaking research, no stone has been left unturned. Over the decades he has amassed thousands of e-mails, pictures, and government documents. He also returned to Vietnam in 1995, tracking down leads and interviewing potential witnesses.

Though Luck had accumulated copious amounts of information over the decades, Kelly's story had never been consolidated into a single, organized narrative. I humbly asked if I could write a story on his brother for publication in a periodical, and thankfully he agreed. Little did I know I was about to embark on the most ambitious writing project of my life.

Luck put me in contact with several of Kelly's USNA classmates as well as surviving members of VA-35 from the 1966-1967 WESTPAC cruise on *Enterprise*, including Red McDaniel. One academy classmate he put me in touch with was Pete Carrothers, who had been serving as a Reconnaissance Attack Navigator with Heavy Reconnaissance Squadron Seven (RVAH-7) *Peacemakers* on *Enterprise* when Red and Kelly were shot down. In conversing with Pete I became interested in and learned quite a bit about the role of the Vigilante in Vietnam. Perhaps sensing my intrigue, Pete put me in contact with many surviving members of his squadron. That, in turn, led to writing an accompanying story on RVAH-7.

It took two years of research, interviews, and writing to complete the two stories. However, by that time my efforts had produced works well beyond the size constraints imposed by most periodicals. I decided instead to write a book about Navy combat squadrons in the Vietnam theatre. As all of my VA-35 and RVAH-7 contacts had been on the 1966-1967 WESTPAC I decided to focus my efforts on that pivotal and consequential deployment. However, the relationships I had established did not extend far beyond those two squadrons. I realized I would need to cast a wide net across the air wing to find enough men willing to share their stories.

One of the other combat squadrons in the *Enterprise* air wing that cruise was the light attack squadron VA-113 *Stingers*, which flew the A-4 Skyhawk. I remembered that a website I often use for research, *Rolling Thunder Remembered*, was created and managed by retired Admiral Jeremy "Bear" Taylor, who had flown two tours with VA-113, including the 1966-1967 WESTPAC. I reached out to Bear, and he generously agreed to be a part of my writing effort. He also sent out an e-mail to surviving members of the squadron asking if they would like to be a part of my book. Numerous men replied in the affirmative, including retired Admirals William "Bill" Bowes and Ernest "Ernie" Christensen (a man whom Taylor described as "The bravest of the brave"). I also received invaluable assistance from VA-113 pilot Jay Greene, who generously shared with me the daily diaries he kept during his two combat deployments. Without those diaries, I could not have written about the *Stingers* in such detail.

While communicating with Bill Bowes I mentioned to him that I needed a contact within F-4 Phantom squadrons VF-92 *Silver Kings* or VF-96 *Fighting Falcons* for my fourth chapter. He replied that he knew retired Commander Terry Born, a highly experienced and aggressive F-4 pilot who had served three combat tours in Vietnam, including a 1966-1967 tour with VF-96 on *Enterprise*. I contacted Terry, who not only agreed to help but also put me in contact with several surviving members of VF-96 from that deployment. Many of them also wholeheartedly agreed to assist in my endeavor.

It took a further two years of work to complete the subsequent two chapters. During one of my follow-up conversations with Ernie Christensen, he told me the story of his friend and fellow Blue Angel pilot Commander Harley Hall, shot down and lost on the last day of the conflict 27 January 1973. Ernie had also been flying a combat mission that afternoon and witnessed first-hand the loss of Hall. With the assistance of the *Forward Air Controllers Association* I located and contacted retired Air Force Captain Adam West, who was the FAC who had been working with Hall that day. He also witnessed the tragic events unfold and was eager to share his memories. This resulted in a fifth chapter.

In early summer 2022, with my manuscript nearly complete, I decided to write an additional chapter on an under-appreciated aircraft vital to Navy combat operations of the era, one that saved hundreds of lives—the A-3 Skywarrior. I contacted the *A-3 Skywarrior Association* who generously put me in contact with several Vietnam-era A-3 pilots and NFOs. Unfortunately, none of those men had been in the air wing aboard *Enterprise* during the 1966-1967 WESTPAC. However, the majority of them had flown missions in 1972-1973, during the pivotal *Linebacker* and *Linebacker II* operations that concluded the conflict.

After any endeavor of such scope and breadth as *Across the Wing*, there are many people to thank. First is Larry Liguori, who originally told me about Kelly Patterson. Then there is Luck Patterson, who not only inspired the book but also encouraged and assisted me during the four years it took to write. A huge debt of gratitude also goes to Jeff Erickson, Senior Technical Fellow Boeing Corporation (ret.), and Mike Nishina, both fellow Lyon Air Museum Docents. I am also grateful to former A-6 B/N Peter Adams Young, author of *One Hundred Stingers*, for his thorough proofreading. I would also like to thank Chris Hobson, author of the book *Vietnam Air Losses*, and former F-4 fighter pilot and retired Navy Captain Dave Lovelady, who brought that seminal work to the internet. Through their efforts they have preserved the names and memories of those who served, and those who never came home. The majority of loss accounts other than those provided to me first-hand come from the records of *Vietnam Air Losses*. Lastly, eternal thanks go to my wonderful wife Tamara, who wholeheartedly and unwaveringly supported my effort.

Though I have strived to ensure historical accuracy in *Across the Wing* the passage of many decades, varying recollections, and differing records made this a challenge at times. I welcome all feedback from readers and will do my best to respond in a timely manner.

Today the skies over Southeast Asia are calm; Yankee Station is no more than a distant memory—an otherwise unremarkable patch of water in a vast ocean. The lush jungle landscapes, soaring karst mountains, winding rivers, and shimmering shores where hundreds of thousands of Americans fought, and tens of thousands died, remain quiet. Yet, half a century later the reverberations of that conflict so long ago still echo upon the land, sea, and sky. The battles fought will forevermore live in the hearts and minds of those who served and in the hallowed and treasured memories of those souls who flew into the unknown, giving their lives defending the country they would never set eyes upon again.

Dan Heller
acrossthewing@protonmail.com
May 2023

# Betrayed

"Nearly twenty years later, I saw former Secretary of State Dean Rusk being interviewed by Peter Arnett on a CBC documentary called 'The Ten Thousand Day War.' Mr. Arnett asked, 'It has been rumored that the United States provided the North Vietnamese government the names of the targets that would be bombed the following day. Is there any truth to that allegation?'"

"To my astonishment and absolute disgust, the former Secretary responded, 'Yes. We didn't want to harm the North Vietnamese people, so we passed the targets to the Swiss embassy in Washington with instructions to pass them to the NVN government through their embassy in Hanoi.'"

Lieutenant General John "Pete" Piotrowski, USAF (ret.)
*Basic Airman to General: The Secret War & Other Conflicts: Lessons in Life and Leadership*
Xlibris Publishing 2014

# Chapter I

# Ray Gun 502 is Burning

Heavy machine gun emplacements, sandbags, and barbed wire lined the perimeter at the headquarters of the 1st Battalion of the 1st Marine Regiment (1/1) west of Hoi An, Quang Nam Province, Republic of South Vietnam. It was April 1967 and inside the fortified base, a young, tall, and lanky USMC 2nd lieutenant prepared his platoon of 35 men for a three-day sweep of the area for suspected VC positions. As the platoon received the briefing in a hooch a fit, young man wearing Navy fatigues with silver lieutenant bars on his collars and gold flight wings on his chest appeared. Hauling a green duffel bag, he returned the crisp salutes of numerous enlisted personnel manning the base near the ancient city.

The Marine officer instantly recognized the man who, aside from the Navy chaplain, physicians, corpsmen, and other medical support personnel attached to the battalion, seemed somewhat out of place in the Marine forward firebase. Approaching each other no words were necessary as wide grins appeared and handshakes were exchanged. Two brothers, one looking for a fight with the enemy and the other on leave from his combat tour, reunited in a theatre of war some 8,000 miles away from home. However, with a three-day patrol lingering it created a problem. The Marine's older brother was a naval aviator, not an infantryman trained for jungle warfare. The U.S. government had invested considerable amounts of money in his education and training, and not for him to be on the ground slugging it out with VC guerrillas. However, it was likely the only chance the two brothers would have to see each other during their respective tours. The choice for them became clear.

Obtaining permission from his superiors, the Marine took his older brother to supply where a flak jacket, helmet, and M-16 with ammunition were issued to him. With his lieutenant bars and gold wings removed the Navy flyer headed into the bush with his younger brother. For two days they patrolled sandy, open, and sometimes jungled coastal scrubland broken by tree line-hidden hamlets full of nearly impenetrable bamboo hedgerows. The patrol proved uneventful, which provided the brothers some time to reminisce and talk about family and friends back home. Despite the circumstances, it proved to be a special time between them, sole siblings who once shared a bedroom as boys now serving together in Vietnam as men.

After two days on patrol, the Marine platoon was called to a suspected VC hotspot in the foothills of the Annamite Mountains. As a CH-46 Sea Knight helicopter landed and the platoon began boarding the two brothers reluctantly said their farewells. It had been a good couple of days, and their spirits were bolstered. In a jungle clearing west of Hoi An the Marine scrambled to a porthole of the ascending helicopter to see his older brother climb aboard an amtrac and begin the journey back to his seaborne command. A moment frozen in time, one that neither brother would ever forget.

Kelly and his younger brother Luck in Hoi An April 1967.

One month later, 19 May 1967, a hazy Friday morning. The darkness of dawn quietly gave way to daybreak as the churning waters of the Tonkin Gulf began to shimmer in rays of bright sunshine through the broken cloud layers. Off the coast of North Vietnam, southwest of China's Hainan Island, the faint outlines of three massive vessels began to appear. United States Navy aircraft carriers USS *Enterprise* (CVA(N)-65), USS *Kitty Hawk* (CVA-63) and USS *Bon Homme Richard* (CVA-31) were operating from Yankee Station as part of CTF-77, performing round-the-clock missions deep into North Vietnam.

In IOIC[1] onboard *Enterprise* the pace of activity was rapidly picking up. Aircrew, intelligence officers, meteorologists, and other specialists from the air wing gathered and reviewed photographs and other intelligence on their assigned targets for the late morning strike. Commander Herman "Herm" Turk of VA-35 methodically worked his tired eyes over a map of North Vietnam, covered with colored strips of paper marking ingress routes and locations of known enemy positions that could pose a threat to the strike group.

The *Black Panthers* were one of the Navy's most famous, and storied combat squadrons. Activated in 1934 as VB-3B in Norfolk, in April 1942 they embarked on USS *Enterprise* (CV-6), escorting USS *Hornet* (CV-8) when it sailed towards Japan with Colonel Jimmy Doolittle, his raiders, and 16 B-25 Mitchell bombers onboard. Two months later, while deployed on USS *Yorktown* (CV-5) and under the command of Lieutenant Commander Max Leslie, the squadron played a pivotal role in defeating the Japanese at the Battle of Midway. Flying the SBD Dauntless they were part of the attacks that resulted in the sinking of the Japanese fleet carriers *Soryu* and *Hiryu*, sealing a historic and consequential victory for the United States. The squadron continued to

4

serve in the Pacific, participating in the Solomon Islands and Leyte campaigns. In February 1945 VB-3 took part in the first carrier-based air strikes on Tokyo, followed by support for Marines fighting the vicious and consequential battle for Iwo Jima. The 1966-1967 deployment aboard *Enterprise* was the squadron's first to Vietnam.

VA-35 was part of a strike group going "downtown" as American airmen called the Hanoi area, in a daylight Alpha strike. The targets for *Enterprise* and *Kitty Hawk* strike groups were the Van Dien Truck Maintenance Depot and SAM storage facility just six miles south of the North Vietnamese capital in an area known as "Little Detroit" for the numerous industrial complexes and factories located there. Due to the proximity to Hanoi, the targets were heavily defended by AAA, SAMs, and VPAF Soviet and Chinese-supplied MiG fighters.

The Hanoi TPP was the target of aircraft from which *Bon Homme Richard*, also known as *"Bonnie Dick"* would be using Walleye bombs for surgical precision in the densely populated capital. The Walleye was one of the first precision-guided munitions developed. It made its debut in the Vietnam theatre on 11 March 1967 in a mission against the Sam Son NVA barracks by A-4s of VA-212 *Rampant Raiders* from *Bon Homme Richard*.

Alpha strike targets were designated by the JCS through the DoD as approved by President Lyndon Johnson personally. Navy strike groups typically consisted of approximately 12-18 aircraft from each carrier. Normally composed of both fighter and attack squadrons, each aircraft type was assigned a specific role in the strike. Thus, extensive planning and coordination between multiple command layers was required.

For the Alpha strike on Van Dien, Commander Turk would be commanding aircraft from CVW-9 on *Enterprise* as flight leader of six state-of-the-art A-6A Intruders call sign "Ray Gun". The A-6, also known affectionately as "Drumstick" or "Iron Tadpole," had a side-by-side crew configuration, bulbous radome nose, and oddly protruding refueling probe which gave the aircraft an outwardly awkward appearance.

The Intruder was designed and built by Grumman, a company long revered for engineering some of the most rugged and effective aircraft in Naval Aviation history. The historic F-4F Wildcat, TBF Avenger, and F-6F Hellcat are legends of WWII that played a key role in the Allied victory in the Pacific theatre. Five years later the F-7F Tigercat and jet-powered F-9F Panther proved their worth over the Korean peninsula during the 1950-1953 war.

The A-6 was seemingly designed for combat in the Vietnam theatre. It featured a revolutionary computerized combat, weapons, and navigational flight management system known as DIANE, or Digital Integrated Attack and Navigational Equipment. It was this technology that made the A-6 stand out among its predecessors and contemporaries. It could fly missions day or night in all weather conditions, guided by an INS, advanced Doppler radar, and other sensor systems. These features were very important in Southeast Asia (SEA) where low cloud cover and fog, soaring karst formations,[2] deep valleys, and months of rainy monsoons often made low-level missions impossible for other aircraft. The Intruder could also carry 18,000 pounds of payload and pull heavy g-forces due to the rugged design of the airframe. These capabilities resulted in the Intruder quickly developing a reputation as the de facto aircraft of choice for difficult missions.

Another advantage of the A-6 was the unique crew configuration. During the design phase of the aircraft in the late 1950s engineers at Grumman theorized that a two-man crew sitting side-by-side would work better together and achieve greater results than the typical front-and-back crew configuration of contemporary combat aircraft. The idea was drawn from the effectiveness of RAF Mosquito crews in World War II. It became known as "crew concept" and it worked better than the designers had ever planned.

Not only was the right-seater in an Intruder a navigator, he was also a bombardier (B/N). In addition to skillfully guiding the pilot to and from the target, the B/N also played a crucial role in weapons targeting and delivery. Sitting side-by-side in an Intruder, the pilot and B/N were equal, wholly dependent on each other for mission success or failure, life or death. For the crew concept of the A-6 to work efficiently, the two men had to operate as one. Their relationship had to extend beyond the close confines of the cockpit; they truly had to respect and trust each other for maximum combat effectiveness. It often took several crew rotations before squadron commanders found the right combination of men to fly in combat together. Once a team, the pilot and B/N operated seamlessly, each always knowing what the other was doing. Vocal communication was not always necessary as the crew could communicate simply by glance or gesture. Though the pilot and B/N seats were side-by-side the right seat of the B/N was slightly lower and behind the left-hand pilot seat. This was to ensure the pilot had good visibility out of the starboard side.

Revolutionary for the time, A-6 pilot and B/N each had CRT screens. These displays showed DIANE navigation and target information by the use of radar-generated terrain displays supplied by the advanced radar and INS systems onboard. The B/N also had a typewriter keyboard to interface with the sophisticated computers which powered the system, allowing easy entry of navigation waypoint coordinates. Navigational guidance from the B/N was fed directly to the pilot via a computer link to his own television screen, known as the Vertical Display Indicator (VDI).

Ray Gun 502 BuNo. 152594 at NAS Alameda fall 1966.

The technological advancements of the Intruder required constant maintenance and support. When first deployed to SEA a slew of company representatives from Grumman, Litton, Norden, and other firms accompanied the aircraft aboard the carriers, with a line of communication to engineers in the States. There were so many reps that berthing space became an issue, resulting in some of these civilians sleeping on cots in the hangar deck. Though many early issues with the aircraft were resolved, reps continued to accompany the aircraft during combat deployments to the theatre.

Bill Schultz was a Field Service Weapons Systems Integrator for Grumman and team leader of five civilian technicians aboard *Enterprise*. He was with the A-6 program from the start and saw the airplane through its life of service, "from womb to tomb". Deploying with VA-35 for the squadron's 1966-1967 cruise he recalls:

> By the time VA-35 went to sea in November 1966 we had made many improvements in the aircraft, such as air conditioning for the computer and fewer wiring/mating connector problems. We also benefited from having experienced Navy technicians from earlier deployments who were becoming familiar with the avionics. As a result, by that fall deployment, the reliability of the aircraft had greatly improved, as had the confidence in its abilities. That is really when the A-6 came into its own.

Praise and admiration for the Intruder was high among the men who flew them in combat. Beneath the modern cutting-edge digital avionics was classic Grumman—a tough, rugged, and resilient airplane engineered to survive the rigors of Naval Aviation combat and protect the crew. The two Pratt & Whitney J52-P8B engines provided plenty of power and were highly responsive, critical in situations such as outrunning enemy AAA or in case of a "bolter" landing. It was also forgiving in slow flight such as approaching the carrier for landing, giving ample warning of an impending stall. Visibility from the cockpit was excellent, providing an unobstructed and wider field of view than most other carrier-based attack or fighter aircraft. Highly maneuverable and agile, it could out-turn most of its contemporaries, including the vaunted F-4 Phantom. Critical components such as engines and hydraulics were protected by three-quarter-inch armor plating. In the effusive words of VA-35 pilot Dave Cable, "The 'Grumman Iron Works' beast could take a licking and still get us home. I loved the old bird!"

Onboard *Enterprise* that morning, as squadron commanders planned the mission in IOIC, phones began ringing in staterooms with wake-up calls from squadron duty officers for crews flying the late morning Alpha strike. One of the phones that rang was in the stateroom of 26-year-old Lieutenant James Kelly Patterson, a B/N of VA-35. Outgoing and handsome with brown hair and a stout, full frame his hazel eyes exuded confidence and character. A 1963 graduate of USNA, Kelly had dreamed of being a pilot his entire life. Growing up in South Pasadena, a suburb of Los Angeles, he spent many hours building model military aircraft in exacting detail, which hung from the ceiling of the bedroom he shared with his younger brother George, also known as "Luck". He also built flying models with doped wings which he and Luck would fly in the vast, empty parking lot of the Rose Bowl. Luck remembers:

> Some of the model aircraft would climb in a circular pattern, then fly away to never be seen again. It happened often enough for Kelly to put a note in the aircraft with his name and phone number in case they were found. When he lost a model, which was often, he would simply build another.

Kelly had applied for entrance to USNA during his senior year at South Pasadena High School. His Father, a career Navy yeoman chief who assisted in the creation of NATO in post-World War II Europe, was especially hopeful of an appointment. However, Kelly's initial application was rejected. Determined to become a naval aviator, he spent the 1958-1959 school year at Pasadena City College (PCC), taking advanced classes and earning high academic marks. He re-applied to USNA and was accepted for the class of 1963, arriving in Annapolis in July of 1959 as a plebe with dreams of boundless, blue skies. Academy classmate Kent Maxfield states:

> We were classmates at PCC during the 1958-1959 school year. During that year Kelly and I had a mutual intense desire to attend the academy and, as a result, became good friends. He was a great guy, always upbeat, focused on the task at hand, with a great and very supportive family. Both of his parents were very proud of him and his younger brother. When we were accepted to the academy we made the trip to Annapolis together on one of the earliest commercial jets—a Boeing 707. It was an experience that was a highlight of our young lives.

Kelly excelled at the academy and was regarded by his classmates and superiors as dedicated and professional. One of his roommates, Jim Ring, recalls how Kelly's artistic talent brought them some much-needed relief from the ritualistic hazing inflicted upon the first-year plebes from upperclassmen:

> Football was a big sport at USNA and we learned the week before the season opener that all plebe rooms were required to develop a poster and hang it on our door. The poster was to feature something about how we were going to beat our football opponent that week.

To our surprise, we found out that Kelly was a talented artist who had a knack for developing great ideas to be used on the poster. For fear of running into an upperclassman, we didn't roam the halls to find out what the other plebes in our company had drawn. So, we were pleased to learn that for the first week, we had won. This meant the poster would be exhibited in the rotunda of Bancroft Hall with entries from the 23 other cadet companies for all visitors to see. More importantly, winning meant that our room was relieved of plebe duties for one day the following week. The next week Kelly came up with another great idea and made it into a cartoon, and we won again. This continued week after week, much to the chagrin of the upperclassmen who didn't like that the same room kept winning and getting excused from a day of duty!

Another classmate, Lee Cargill, remembers:

Kelly was a great guy and a wonderful roommate. I fondly remember our years together at the academy, and that he especially liked the smell of rain in Annapolis. He would open the window and take it all in. One spring break Kelly and I stayed at Bancroft Hall and spent our time making a very large glider. I can't remember where we got the plans or kit or whatever, but it was a big one. Didn't fly worth a darn! But it was fun nevertheless.

He and I were very much alike in that the reason each of us decided to come to USNA was to become a pilot. That was our dream, to graduate from the academy and fly for the Navy.

Kelly was so ambitious to become a naval aviator that he seriously considered leaving USNA after his plebe year to enlist as a NavCad, which would fast-track him to flight training. However, his Father insisted he stay at USNA and graduate, which would bode better for his Navy career. Kelly's brother Luck remembers an emotional conversation between his Father and brother on this subject in their California home, where their Father prevailed. Though Kelly was disappointed, his desire to fly remained. "I don't think Kelly cared about advancing through the officer ranks of the Navy insomuch as he just wanted to fly" remarks Luck. After this fateful conversation, Kelly returned to the academy, putting his aviation ambitions on hold. For the time being.

During his third year at the academy, a certain plebe would seek refuge in the room of Kelly and his roommates, who did not particularly enjoy hazing the lowly first-year cadets as other upperclassmen did. Jim Ring recalls:

> He would knock on our door and request to come in and spend time in our room before the evening meal, just relaxing. We gladly granted his request, just asking him to be quiet so we could study. NCAA rules did not allow freshmen to play varsity sports, so we did not know just how good of an athlete Roger Staubach was until the following year.

USNA Graduation 1963.

Receiving his wings 1966.

Shortly before graduation from USNA in 1963, the senior midshipmen underwent an extensive physical examination to determine eligibility for their chosen path in the Navy or Marine Corps. Kelly had spent five years studying for his undergraduate degree, first at PCC and then at USNA. Countless hours had been spent under dim lights, intensely focused on a myriad of finely-printed textbooks and notepads. Unknowingly, this had taken a toll on Kelly's eyesight, causing him to develop myopia. This condition disqualified him from being a pilot, and he was rejected during the physical exam. To say he was emotionally crushed would be an understatement. His lifelong goal, his dream, had been taken away with the cold signature of a Navy physician. Lee Cargill was there that day:

> I remember well the day that we took our physical exams and Kelly learning that he had been disqualified for flight training because of his eyesight. He had tears in his eyes. Today I am positive he would have eventually received a waiver, especially with his motivation.

Jim Ring also witnessed the aftermath of the crushing blow:

> After Kelly failed the flight physical he became depressed. He used to lie on his bed for long hours with a pillow over his head. There was nothing his friends at the academy could do but encourage him to keep going.

Instead of pilot training, as he had hoped and dreamed, Kelly went to San Diego where he completed Officer Fleet training, before being assigned to the WWII-era destroyer USS *Renshaw* (DD-499). He spent about a year on *Renshaw* but was predictably restless. Undaunted, Kelly was determined to get into the cockpit of a Navy aircraft and fly, one way or the other. Applying for NFO training, he was accepted despite his myopia. Though he would not be piloting an aircraft, he would still be flying and that was better than the monotony of serving as a junior officer on an aging, rusting destroyer.

Kelly attended NFO training from September 1964 to April 1965, followed by specialized navigator training in an RA-5C Vigilante of RAG squadron RVAH-3 *Sea Dragons* from April to June at NAS Jacksonville. As the DIANE system in the A-6 required a high degree of skill, dexterity, and acumen only the top graduates of each training class were eligible to obtain coveted A-6 B/N slots. Intruder crews were known to have, in the words of Vice Admiral William F. Bringle, Commander Seventh Fleet, "An abundance of talent, courage and aggressive leadership".

Kelly was able to meet the lofty requirements and began A-6 B/N training with RAG VA-42 *Green Pawns* at NAS Oceana in Virginia. It was at this time he met and was paired with 35-year-old pilot Lieutenant Commander Eugene "Red" McDaniel. They trained together in the A-6 extensively from June through August, culminating in their assignment to VA-35 *Black Panthers*.

Red McDaniel was a strapping six-foot, three-inch native of Kinston, North Carolina. The son of poor sharecropper parents, he had attended Campbell Junior College on an athletic scholarship for baseball, then went on to Elon College to finish his undergraduate degree. By the time he met Kelly at VA-42, he was already a seasoned naval aviator, having earned his wings in 1956. He had both flown and instructed in the A-1 Skyraider ("Spad" or "Sandy") and made a Mediterranean cruise on

USS *Independence* (CVA-62) before transitioning to jets and ultimately the A-6. Red remembers Kelly fondly:

> A great guy, mild-mannered with a strong moral character. I remember him writing a letter to his parents stating he had no problem with humanitarian aid being given to North Vietnamese civilians. He was also very conscientious of collateral damage and always careful to positively identify the target before releasing ordnance. In fact, at times he was so careful it frustrated me.

Intruder Ball September 1966 NAS Oceana
Back Row: Bill Gaynor, Bob Benjamin, Kelly
Front Row: Jack Farady, Richard Slaasted

Ed Sadowski, one of Kelly's VA-35 squadron mates, also recalls his moral character:

> He was only a year older than me, so we got to be pretty good friends. Early in my first assignment with the Navy, I learned about per diem, which is money for daily food and lodging you received when you went on temporary duty, similar to an expense account. One day I was speaking to Kelly about a two-week assignment he had been on where he was given full per diem reimbursement for lodging since the BOQ at the base he was at was regarded as sub-standard. He told me that he stayed at the BOQ anyway, found it perfectly acceptable, and was planning on paying back the per diem lodging reimbursement to the Navy. I was flabbergasted by his honesty. I had never before met anyone with this sort of integrity. He taught me a life lesson, one I have carried for over fifty years.

Luck remembers that Kelly had trouble returning the reimbursement money to the government, as there was apparently no protocol for doing so. He eventually wrote a personal check to the U.S. Treasury and mailed it off to Washington, DC.

In early autumn 1966 the men of VA-35 began preparing for their impending eight-month deployment to Vietnam aboard *Enterprise*, scheduled to begin in November 1966 and last through July 1967. Red recalls that with *Enterprise* ported across the country at NAS Alameda the squadron began moving from Oceana to the West Coast in a prelude to their deployment:

When we were flying from Oceana to Alameda for deployment aboard *Enterprise* we had some fun. A little while after we took off Kelly and I switched seats in flight. Now, the cockpit of the A-6 is cramped to begin with, and I am a tall individual. I don't remember exactly how we did it, we must have twisted ourselves into pretzels. But we did it, and Kelly sat in the left seat flying for quite a while. If our command had found out we would have received a few choice words from the skipper, at the least! Fortunately, Kelly was no stranger to the control stick of an aircraft, having taken glider lessons in high school and earning a private pilot license while in Oceana.

The act of switching seats in flight was symbolic of just how highly the two men regarded each other. As aircraft commander Red was responsible for the Intruder as well as the lives of the crew aboard. To let an NFO sit in the left seat and take control without a very good reason was strictly forbidden by military regulations. It was also viewed by many pilots as an affront to the years of grueling training they go through to earn their wings. However, the act was far from one of willful disregard or defiance. Instead, it was a mutual act of faith and trust, representative of the professional and personal relationship the two men were developing, one which cemented their bond as a team. It also fulfilled Kelly's dream of flying a Navy aircraft, albeit for just a few short hours. It was during this flight across the country that the two men, for the first time, candidly talked at length about what awaited them in the skies over SEA, first tours to the theatre for both of them. Information gleaned from returning aviators had not been encouraging.

Kelly standing proudly atop an A-6.

Kelly and Red on Yankee Station.

VA-35 *Black Panthers* aboard *Enterprise* fall 1966

Back Row: Duff, Ellison, Sadowski, Slaasted, Fardy, Gaynor, McDaniel, Ross, Baric, Turk, Van Lue, Foote

Front Row: Unknown, Gordon, Cozzi, Johnson, Patterson, Leonard, Borchers, Carpenter, Mallek, Cable, Bremner

The conflict in Vietnam had rapidly escalated since the Tonkin Gulf Incident nearly three years prior, with more than half a million troops in the theatre compared to a paltry 23,000 in 1964. During this period of escalation, it had become painfully clear to the United States that North Vietnamese forces were not untrained ragtags armed with obsolete hardware left over from French colonial rule. They were a tenacious, formidable opponent well-versed in jungle, guerrilla, and unconventional warfare, supplied with state-of-the-art military hardware, as well as trainers and advisors from their communist allies the Soviet Union and China. The VPAF had also become a force to be reckoned with. In just three short years it grew from a handful of aged piston aircraft to squadrons of Soviet and Chinese-supplied MiG fighter aircraft. The VPAF pilots who flew these aircraft had been trained by their communist benefactors and some were highly capable combat aviators.

North Vietnam also had a civilian workforce of tens of thousands who could repair bombed roads and bridges virtually overnight. Bomb craters on roads would be filled and the road usable again within hours. A bridge bombed and rendered inoperable one day would be partially operational again within a day or two. Due to this American aircrews often found themselves attacking the same target numerous times. The Long Bien ("Paul Doumer") Bridge running east and west over the Red River between Haiphong and Hanoi and the Ham Rong ("Dragon's Jaw") Bridge over the Song Ma River near Thanh Hoa are prime examples of their extraordinary rebuilding efforts. Dragon's Jaw, completed in 1964, spanned 540 feet and was supported by nine concrete piers and abutments 40 feet thick, which anchored the bridge at both ends. The Air Force and Navy flew a staggering 873 sorties against the bridge between the years 1965-1972.

It was not just military factors fueling the problems on the American side. There were many political issues with the way the conflict was being run half a world away in Washington, DC. The micro-managing and restrictive ROEs forced upon combat aircrews by the administration of President Lyndon Johnson were costing men their lives and having a terrible effect on morale. Enforcement of the ROEs was strict and violation often resulted in administrative punishment or court martial.

North Vietnamese airfields not on a target list were off-limits. VPAF aircraft could only be fired on if they were airborne and showed hostile intent. AAA and SAM sites could not be attacked unless they engaged first. Any air defense site under construction could not be attacked, as foreign technicians may be present. Military trucks could not be attacked unless they were on a road and displayed "hostile" intent. Very few enemy elements could be attacked unless they were on the approved target list, initiated hostile action first, or special permission was received. So restrictive were the ROEs that days after his January 1965 inauguration President Johnson, referring to American airmen flying combat missions in Vietnam, boasted "They can't even bomb an outhouse without my approval!"

Perhaps as frustrating were the restrictions on severing the North Vietnamese supply lines. Sixty miles east of Hanoi is the deep-water port city of Haiphong, where North Vietnam received approximately 80% of its war materials from the Soviet Union and China. Here the ships were unloaded and weapons were dispersed by truck, rail, and waterway to positions around the north. These weapons also made their way to South Vietnam via the Mu Gia and Ban Karai passes, Laotian access points to the Ho Chi Minh Trail. In reality, the Ho Chi Minh trail was not one, but a series of primitive roads and paths that meandered down the panhandle of North Vietnam, winding in and out of Laos and

Cambodia as it made its way into South Vietnam before turning east. These transit systems included rest stations, field hospitals, communications centers, and bunkers, enabling streamlined logistical coordination of supplies and fighters flowing south. Once on the trail under triple jungle canopy, the vehicles used to transport these arms became difficult if not impossible for American aircraft to locate and destroy.

Hon Gai (now Ha Long) was another North Vietnamese port where arms and supplies were received. Twenty-five miles northeast of Haiphong, it accounted for a much smaller percentage of shipments from the Soviet Union and China. As with Haiphong, it was repeatedly attacked by American air forces during the conflict however the ROEs in place at the time limited the damage inflicted. It would have made sense politically and militarily to stop the flow of arms at the source by destroying the port facilities and mining the harbor of Haiphong such as what had been recommended to President Johnson in 1965, a recommendation he declined (later embraced by President Nixon during the Easter Offensive of 1972 in *Operation Pocket Money*). However, the omnipresence of numerous Chinese and Soviet vessels raised fears within the Johnson Administration that attacking the port facilities at Haiphong or mining the harbor could result in foreign collateral damage, provoking a Cold War response such as from the Chinese in Korea in 1950. Instead, non-floating maritime mines were laid by Intruders of VA-35 in February 1967 across the Song Ca and Son Giang rivers, the first time maritime mines were utilized during the conflict, and the first time maritime mines were deployed by jet aircraft. This was to prevent smaller vessels such as junks, barges, and sampans from transporting arms inland along the many rivers and tributaries.

Much of Haiphong itself and the main port facilities remained strictly off-limits. Also off-limits were the thirty miles south of the border between North Vietnam and China, which included several railroad lines used to transport military materials between the communist neighbors. The thirty-mile restriction also prevented American fighters from pursuing enemy MiGs to the border.

The North Vietnamese were also aided when President Johnson ordered numerous temporary bombing halts as a sign of goodwill to the North Vietnamese people. These pauses, lasting days to months in duration, did not result in NVN seeking a peaceful resolution to the conflict but instead allowed them to repair, rearm, refresh, and retrain. There is no question the North Vietnamese took strategic advantage of the ROEs and bombing halts, to the great detriment of American military efforts, resulting in additional casualties.

Though Kelly and other aviators strived to keep their personal feelings separate from their duties, it was inevitable that doubts crept into their minds. These doubts certainly seemed justified, not only by the discouraging and demoralizing ROEs but also by the fact that much of their efforts seemed to be in vain, with no clear path to victory. It also concerned and disheartened the Americans fighting the war that public support for the conflict was plunging, with much of the anger being taken out on those serving in uniform. Soldiers, sailors, and airmen were returning home to angry protestors who openly attacked them, verbally and occasionally physically, accusing them of heinous crimes. It was to the point that upon returning to the States those in uniform were advised to not wear them in public. These were topics Kelly discussed often with his fellow aviators such as RA-5C Vigilante RAN and USNA classmate Pete Carrothers of RVAH-7 *Peacemakers*:

Kelly and I were both on *Enterprise* for the WESTPAC cruise of 1966-1967. We spent many an hour together in our staterooms trying to find a rationale for why President Johnson was letting politicians run the war. We could take pictures of a SAM assembly site outside Hanoi, but it couldn't be attacked as there were likely Soviet technicians there. The bicycle assembly plant two blocks down the street was okay to bomb. The North Vietnamese were not shooting surface-to-air bicycles at our aircraft, but they were sending up plenty of SAMs. Kelly, myself, and others were extremely frustrated with these restrictions.

Lee Cargill agrees. "When I think of the type of tactics that were used in the early years of the war, they were extremely poor and led to losses that were way too high." Frustrations mounted as American losses accumulated and North Vietnamese combat effectiveness increased, especially in the size and skill of their rapidly growing air defense network.

The greatest threat to American aircraft over North Vietnam came not from MIGs, who largely employed hit-and-run tactics, but rather ground fire. The most advanced ground-based air defense weapon in the North Vietnamese inventory was the Soviet S-75 *Dvina* (SA-2) air defense system which utilized the V750 SAM missile, which measured 35 feet in length and could travel in excess of Mach 3. American aircrews described the missile as a "flying telephone poll with fins". The system used two radar types, typically mounted atop trucks for ease of mobility. "Spoon Rest" search radar provided initial acquisition before the target was passed to "Fan Song" radar.

Christmas 1966 in VA-35 ready room.

Happy times below deck.

SA-2 missiles did not have independent guidance. They were wholly reliant on a ground operator who guided the SAM to its target. If taken one-on-one pilots could often out-maneuver a SAM by breaking hard towards the missile at the right time. The SAM, traveling at supersonic speed, would be unable to turn and subsequently detonate. A SAM was easier to outmaneuver if the pilot was able to spot it shortly after launch when the solid-fuel first-stage booster rocket emitted a large puff of white smoke when it separated. However, if the ground was obscured by clouds, haze, or smoke the missile often could not be visually spotted until it emerged from the obscurations. This gave pilots just precious seconds to react.

Even though SAMs were a formidable threat, Soviet and Chinese AAA with sizes ranging from 12.7mm to 100mm were responsible for the overwhelming majority of American aircraft lost to hostile ground fire during the conflict. Both 57mm and 85mm AAA guns could be radar-guided by "Fire Can" or "Flap Wheel" radar systems, making their accuracy deadly. Relatively inexpensive and easy to transport and operate, large numbers of these AAA guns ringed high-value targets such as POL factories and storage facilities, power plants, and bridges. Approximately 85% of American aircraft combat losses during the conflict came from ground-based AAA which included small arms such as rifles and machine guns. These handheld weapons could damage or disable even the most sophisticated and costly of combat aircraft.

As ingress and egress routes of aircraft on Yankee Station remained mostly static, the North Vietnamese smartly lined those routes with these highly effective weapons. They often positioned them on steep mountain cliffs and karst formations to better their aim. They accomplished this not only by mechanization but also by sheer manpower and even animals ranging from oxen and water buffalo to elephants.

They also slyly hid AAA guns on rails inside mountain and limestone karst caves, where they could be wheeled out and fired then hidden again before being located and destroyed. Though the subsonic Intruder was agile and nimble, it was not particularly fast when compared to the F-8 Crusader, F-4 Phantom, or RA-5C Vigilante, aircraft that could reach supersonic speed in seconds. That made SAMs and AAA all the more dangerous for A-6 crews.

After a particularly difficult mission over Nam Dinh in February Kelly remarked to Red that it seemed like the North Vietnamese had fired more ordnance at their aircraft than they had dropped on the target. After landing back on *Enterprise* they were amazed and relieved they had made it out unscathed. Pete Carrothers remembers:

> The NVA had a major presence in Nam Dinh, including barracks and training facilities. I recall that it was just as heavily defended as Hanoi and Haiphong. We [RVAH-7] caught some flak in the fuselage over Nam Dinh which is pretty remarkable. On a photo run with only an F-4 escort, we never went less than Mach 1.2 (913mph) at 3,500 feet, which doesn't give the AAA much time to get a lock. We were surprised to see just how much 37mm and 85mm AAA there was.

The mission to Nam Dinh was so notable it was featured on the front page of the *New York Times* on 21 February 1967. Both Red and Kelly were quoted in the story. A memorable passage from the article is Kelly stating "McDaniel pants like a puppy dog every time we get near the place. I guess you'd say it's a tough target".

Kelly aboard *Enterprise* in Pearl Harbor en route to Yankee Station.

Richard Slaasted (leaning on cowling), Stu Johnson, Dave Cable and Red McDaniel.
Flight deck of *Enterprise* Yankee Station.

As late morning approached, the men of VA-35 assigned to the strike assembled in their ready room for a briefing from Commander Turk on details of the mission and from a squadron AIO on areas of SAM and AAA concentrations. The weather at their target was forecast as broken clouds, normally adequate for Intruders to acquire the required visual confirmation of their objectives.

Flying in the right seat next to Turk would be B/N Lieutenant (junior grade) Keith Urbanek. They were the lead aircraft, flying with the call sign "Ray Gun 1." "Ray Gun 2" was Ensign Nick Carpenter and B/N Ensign Richard Slaasted. "Ray Gun 3" was Lieutenant Commander Red McDaniel and Lieutenant Kelly Patterson. "Ray Gun 4" was Lieutenant Steve Owen and B/N Lieutenant (junior grade) Bruce Borchers. "Ray Gun 5" was Lieutenant Commander Bob Miles and B/N Lieutenant (junior grade) Ken Van Lue. "Ray Gun 6" was Lieutenant (junior grade) Dave Cable and B/N Lieutenant (junior grade) Stuart Johnson.

The Intruders from *Enterprise* would each be carrying 22 Mk.82 bombs; five and a half tons of high explosives. Also in *Enterprise* compliment would be approximately a dozen A-4 Skyhawks of VA-56 *Champions* and VA-113 *Stingers*, each carrying about 4,000 pounds of Mk.82 bombs. Rounding out the *Enterprise* strike group were approximately six F-4 Phantoms of VF-96 *Fighting Falcons* call sign "Showtime" and VF-92 *Silver Kings* call sign "Silver Kite" which were equipped with Mk.82 bombs for flak suppression to take out any threatening air defenses as well as AIM-9 Sidewinder and AIM-7 Sparrow air-to-air missiles for any MiGs that may appear. F-8E Crusaders of VF-211 *Fighting Checkmates* and VF-24 *Red Checkertails* from *Bon Homme Richard* would be providing CAP fighter cover for the Skyhawks and Phantoms attacking the Hanoi TPP.

Red and Kelly pre-flighted their A-6, marked just below the cockpit as side number 502 (BuNo. 152594). After flying over eighty combat missions together, it had become a familiar and streamlined routine for them. They carefully inspected all the control surfaces, fuselage, ordnance, landing gear, and tires. They then checked their torso harnesses, escape kits, and the standard-issue .38 caliber revolvers with extra ammunition kept in bandoliers across their chests. Many airmen were also keen to carry heavier sidearms as their pistol was no match for long rifles used by the NVA and local militias.

The ejection escape kit contained first-aid supplies such as bandages, antiseptic, and doses of morphine. Smoke canisters, flares, a flashlight, and a mirror were also included to signal rescue forces, along with a compass, map, "blood chit,"[3] insect repellent, mosquito netting, and other items. Along with the escape kit the aviators carried a portable PRC-63 transceiver radio with alert beacons ("beeper") that could be picked up and located from the air. Perhaps as important for survival were the canteen and bottles of drinking water carried in various pockets of their flight suits.

During their topside pre-flight preparations Captain Holloway, skipper of *Enterprise*, stopped by. He often walked the flight deck before aircraft launch, giving words of encouragement to the men and wishing them good hunting. It was also likely he wanted to see off VA-35 personally, having served with the squadron while deployed on the carrier USS *Kearsarge* (CVA-33) during a 1948 Mediterranean cruise flying the TBM-3W Avenger. In words he surely came to regret Holloway told Red and Kelly that it should be an easy mission, as there had been no reports of heavy flak near the target.

As they climbed into their Intruder the brown-shirt plane captain helped strap Red and Kelly into their GRU-5 ejection seats. They also attached their G-suits to the air valves in the cockpit. G-suits are inflatable air bladders around the legs, thighs, and torso that keep blood flowing to the brain during extreme g-maneuvers, preventing the pooling of blood in the extremities which can starve the brain of oxygenated blood and lead to a deadly black-out condition.

Preparing to launch aircraft, the OOD of *Enterprise* ordered the bow into the wind, providing about 30 knots of headwind down the flight deck. Few things are more wondrous than the deck of an aircraft carrier during flight operations. It is an orchestrated dance; a ballet of people, ordnance, heavy equipment, and aircraft. There are plane captains, fuelers, handlers, ordnance specialists, maintenance personnel, and aircrew all working together in unison, like a well-oiled machine. However, just one second of inattention or complacency can quickly result in injury or death. During the conflict serious incidents on *Enterprise*, USS *Forrestal* (CVA-59) and USS *Oriskany* (CVA-34) cost 206 men their lives and destroyed dozens of aircraft.

With the announcement from "1MC" to "start 'em up" Red and Kelly ran through the engine start checklist while an MD-3A aircraft tractor equipped with a gas-turbine starter unit ("huffer") pumped compressed air into the two Pratt & Whitney engines. They shortly came to life with a loud whine, and several minutes later DIANE came online.

Red and Kelly began running through the pre-takeoff checklist, ensuring their aircraft was ready for flight. Although the air conditioning packs were online, with the engines at idle barely a whisper of cold air came through the vents. The heat in the cockpit was stifling, and familiar beads of sweat began running down their necks, soaking the collars of their flight suits with perspiration.

Kelly, Nick Carpenter, Richard Slaasted and Red McDaniel receiving a briefing in VA-35 ready room.

B/N Bill Gaynor and pilot Ed Leonard.

B/N Stuart Johnson and pilot Dave Cable.

B/N George Mallek and pilot Art Barie preparing for a mission.

Ray Gun 1 and Ray Gun 2 catapulted off the bow with mighty roars, putting Red and Kelly next in line. The plane director, a young yellow-shirt enlisted aviation deckhand, began guiding the Intruder towards the catapult, easing the aircraft into place. The plane director then made fists with his hands and held his arms above his head with his wrists crossed. Red applied the wheel brakes and reduced the engine power to idle, stopping the aircraft. Two green-shirt catapult handlers secured the nose wheel to the catapult shuttle; the JBD rose behind the empennage. The plane director raised both of his arms over his head, fingers from each hand slightly touching, then spread his arms out to shoulder height. Slowly the wings of the A-6 lowered and were locked in place by Red pushing down firmly the wing lock handle located between the seats. With the wings down and locked, wind from the gulf shuddered the aircraft as errant wisps of condensed steam rose from the catapult slot. Manipulating the stick in every direction and pressing down alternatively on the rudder pedals, the control surfaces were responsive and clear. All instruments were indicating properly and within limits. Two red-shirt ordnance handlers scampered underneath the aircraft and removed the bannered steel safety pins from the bombs hanging from the four MER pylons, two on each wing. One of the red shirts held up the pins so Red could count them.

The plane captain then appeared, showing Red three bannered landing gear lock pins. During this time Kelly entered and double-checked map coordinates using the keyboard just under his hooded radar scope. The INS, part of the DIANE system, would keep track of their exact location in the air and assist in guiding them to the target. The plane director then held up closed fists and opened them, signaling Red to release the brakes. The Intruder began to strain against the holdback of the catapult keeping the aircraft stationary and in place.

A green shirt flashed a chalkboard with the weight of the fully loaded aircraft scrawled across it, 60,400 pounds, which Red acknowledged with a thumbs-up, agreeing that the take-off weight matched his calculations. Catapult steam pressure had to be set differently for every type of aircraft and its corresponding weight. Not enough pressure and the aircraft would be unable to gain enough airspeed for takeoff and end up "in the drink". Too much pressure could cause the nose of the aircraft to pitch up prematurely upon launch, or badly damage the nose wheel. Arresting wire (cross-deck pendant) resistance for landing aircraft also had to be set using the same weight considerations.

The young yellow shirt swept his arms towards the catapult officer, transferring control of the aircraft to him. The catapult officer ("shooter"), a non-flying pilot positioned next to the catapult, stood in front of the port wing of Ray Gun 3, his clothing violently flapping in the wind. With Red and Kelly intently watching he raised his arms and twirled his fingers. Red pushed the throttles all the way forward as both he and Kelly instinctively scanned their instruments and annunciator panel to make sure everything was operating properly. Known as "run up," it was their last chance to catch a potential problem before launching.

As 18,000 pounds of thrust hit the JBD, the Intruder pulled mightily against the holdback. With all systems go Red rendered a salute to the shooter then put his eyes forward. At 1020 hours the catapult officer went to one knee under the wing and swept his arm forward towards the bow, touching his hand to the deck. At that same moment the catapult operator, in a protected enclosure next to the flight deck, pressed the catapult release button. Within two seconds Ray Gun 3 was hurtling off the carrier at 143 knots (165 MPH), heading for a rendezvous point near the coast of North Vietnam with other aircraft from *Enterprise*.

Ray Gun 502 getting ready to launch from *Enterprise*.
Two F-4 Phantoms are readying in background.

VA-35 A-6As over the Tonkin Gulf fully loaded with Mk.82 bombs.

Dave Cable and Stu Johnson over the Tonkin Gulf en route to target.

XO Kollman flying Prime Minister of South Vietnam Ky from Saigon to *Enterprise*.

They were joined in the formation by aircraft from *Kitty Hawk*. At this same time, aircraft from *Bon Homme Richard* began their trek to central Hanoi. Seemingly innocuous North Vietnamese sampans and junks, drifting lazily in the gulf, began relaying information to their air defense network inland, who were put on alert. Coastal observers, sighting the armada of aircraft, began ringing their superiors in Hanoi. AAA guns were loaded and SAM radar units were warmed up.

Over the radio USS *Long Beach* (CG(N)-9) call sign "Red Crown," a nuclear-powered guided missile cruiser PIRAZ radar picket ship cruising in the Tonkin Gulf, began warning of enemy aircraft closing fast from the north. Red Crown provided vast radar coverage of the skies over North Vietnam, giving American aircraft another set of eyes. F-8 Crusaders from *Bon Homme Richard* were vectored toward the bogies, which were most likely a formation of VPAF MiG fighters. Once joined in formation the strike group proceeded to the East coast of North Vietnam, going feet dry at Sam Son, southeast of Thanh Hoa. It was then that an oversight in the planning of the mission became apparent. The strike group was having to fly at a greatly reduced speed, much slower than normal for the Intruders and Phantoms. This enabled the slower and heavily laden Skyhawks to climb to their altitude and keep up with the other aircraft taking part in the attack. As the strike group was climbing through 15,000 feet the Intruder airspeeds were so slow that, fully loaded with fuel and weapons, they were near stalling at the slightest movement of the control stick. At such low airspeed and reduced power, they could barely maneuver and thus were highly vulnerable to SAM and AAA fire.

Suddenly, 40 miles from the target, AN/APR-27 RHAW threat detectors in Ray Gun cockpits began blinking. Simultaneously, slow "warble" tones began to blare into their headsets.[4] Turk in Ray Gun 1 came on the radio and announced "blinking red" with other pilots quickly confirming the same. The warble tones then picked up tempo, meaning the Fan Song radar pulse frequency had changed from acquisition to guidance. Numerous missiles had been launched and were in the air heading for the formation, their rising white exhaust plumes intermittently visible through the broken clouds.

An A-4 pilot of VA-113 from *Enterprise*, who had patched a tape cassette recorder into the radio system of his Skyhawk,[5] recorded strike group radio transmissions as they encountered numerous SAMs in the air:

Ray Gun 1: "Blinking red!"

Unknown Aircraft: "We got a blinker."

Ray Gun 1: "Missiles ten o'clock low!"

Ray Gun 1: "Missiles just lifting off at 12 o'clock!"

Ray Gun 3: "Ray Gun 3 breaking up!" (Red's voice)

Ray Gun 4: "Ray Gun 4 is hit!"

Unknown Aircraft: "Missiles coming from 12 o'clock!"

Ray Gun 1: "Ray Gun 4, you say you're hit?"

Ray Gun 4: "Yeah, so is Ray Gun 3. He's burning."

Ray Gun 1: "How you doin' Ray Gun 3?"

Ray Gun 3: Unintelligible voice communication (Kelly's voice)

Ray Gun 3: "OK Red, let's get out of here, huh?" (Kelly's voice)

Unknown Aircraft: "Ray Gun 3 has two good chutes."

> Showtime 7: "Showtime 7, I have two parachutes at 9 o'clock."
> Unknown Silver Kite: "This is Silver Kite. I just saw the plane go in."
> Showtime 3: "Ray Gun, this is Showtime 3, are you going to stay with the chutes?"
> Unknown Aircraft: "We got two beepers."

A single SAM had detonated between Ray Gun 3 and Ray Gun 4, sending a blast of shrapnel into the right side of Ray Gun 3, crippling the hydraulics and rendering the aircraft uncontrollable. Ray Gun 4 was also impacted by the blast concussion, which jarred the aircraft violently. Believing their Intruder to be damaged, pilot Steve Owen and B/N Bruce Borchers set a course for *Enterprise*, where they landed safely a short time later.

Ray Gun 3 began streaming raw jet fuel from the starboard wing and losing altitude. At 2,000 feet above ground level and with airspeed rapidly increasing Kelly ejected, followed a few seconds later by Red. Two good parachutes were observed by other aircraft in the strike group, accompanied by audible tones of locater beacons from the radios attached to their flight suits. Ray Gun 3 continued airborne, unmanned, plunging steeply before becoming fully engulfed in flames and crashing into an area of North Vietnam known as "Banana Valley". Dave Cable, piloting Ray Gun 6 on the mission that day, remembers:

> Within seconds of the missile alert, I saw an orange and yellow flash and felt the concussion of an exploding SAM. It detonated right next to Red and Kelly just after they had jettisoned their munitions and began to take

evasive action. They were at our eleven o'clock position and a little above us, about 200 feet away. As it exploded, shrapnel from the missile sliced right through their aircraft. Their hydraulics were most likely badly damaged, along with other critical systems.

Overhead, F-4 RIO Lieutenant (junior grade) Paul Daley of VF-96, flying with pilot Lieutenant (junior grade) Dick Earnest, took several photos with his Nikon 35mm camera. One photo showed Ray Gun 3 on fire above the jungle floor, streaming jet fuel, while a subsequent photo showed two deployed parachutes drifting toward the ground several hundred feet apart. As Kelly had ejected a few seconds before Red they became separated, landing on either side of a high ridge on a hill. The rest of the strike group continued to their target in Van Dien, only to find a solid overcast layer of clouds at around 8,000 feet. Commander Turk aborted the attack due to inadequate target visibility. Dave Cable is still understandably frustrated that the mission was aborted:

> Having just lost Red and Kelly and after evading multiple SAMs, we were a bit on edge, extremely disappointed, and admittedly quite angry. Even though my B/N had a radar lock on the target, our orders required visual confirmation with our eyeballs. A radar lock was not sufficient for identification. If Turk could not see it with his eyes none of us could, so we aborted. We left the area, departing to the south en route to our secondary target, a bridge on Route 1 a few miles north of Vinh.

After aborting the attack Nick Carpenter and B/N Slaasted broke off from the strike group and headed back to where Ray Gun 3 had gone down. While flying over the crash site they picked up two beepers. Nick Carpenter detailed the exchange in a subsequent official report:

> "This is Ray Gun 502" Nick heard over the radio, immediately recognizing it as Kelly's voice.
> "This is Ray Gun 505, read you loud and clear" Nick replied.
> "Is that you, Steve?" Kelly asked, thinking he was talking to Steve Owen.
> "Negative, this is Nick. What is your condition?"
> "I am okay but I have a badly broken left leg and won't be able to move" Kelly answered.
> "Roger that 502" Nick replied.

Just seconds later another voice came over the radio:

> "This is 502 on deck" called out Red.
> "Roger, I read you" Nick responded, "what is your condition?"
> "I am okay" Red replied, not mentioning his badly injured lower back.

Nick contacted rescue forces on the radio and relayed the coordinates and conditions of Red and Kelly. The area where they landed was hilly and jungled, remote but accessible to rescue forces, with no local enemy activity or hostile ground fire yet observed. With a full bomb load and their A-6 low on fuel, they flew over the men one more time.

Unmanned Ray Gun 502 on fire.

Red and Kelly parachuting into the jungle.

They rocked their wings as they passed overhead, setting a course for feet wet and *Enterprise*. Both men were eager to get back and monitor the rescue operation from the squadron ready room, and if necessary, would launch again to provide support for rescue efforts.

Red and Kelly were not the only naval aviators lost over North Vietnam that day. Pilot Commander Richard Rich and RIO Lieutenant Commander William Stark were flying an F-4B Phantom of VF-96 from *Enterprise* call sign "Showtime 604". Their aircraft was hit by multiple missiles just minutes after Red and Kelly were shot down, causing catastrophic damage. Stark ejected but was seriously injured. He survived and was captured about 20 miles south of Hanoi. Imprisoned as a POW for six years, he was repatriated during *Operation Homecoming*. Rich, who was the XO of VF-96, was ultimately declared KIA. An F-4B (BuNo. 153004) of VF-114 *Aardvarks* call sign "Linfield" from *Kitty Hawk* was also lost to a SAM on the Van Dien mission. Piloted by Lieutenant (junior grade) Joseph "Charlie" Plumb, a 1964 graduate of USNA, and RIO Lieutenant (junior grade) Gareth Anderson, they were hit in the belly of the fuselage by shrapnel from a SAM. The Phantom had begun to disintegrate when the men ejected near Xan La and were quickly captured. Both were released during *Operation Homecoming*. Anderson would later lose his life on 21 June 1976 while piloting an F-14 Tomcat during a training flight from NAS Miramar.

At the same time the strike force from *Enterprise* and *Kitty Hawk* were tangling with SAMs south of Van Dien, aircraft from *Bon Homme Richard* were attacking a Hanoi TPP with Walleye bombs. While two A-4E Skyhawks from VA-212 led by Skipper Commander Homer Smith executed their attack, six F-8 Crusaders from VF-211 and VF-24 provided CAS.

Lieutenant Commander Kay Russell, flying an F-8E (BuNo. 150930) call sign "Nickel 109" of VF-211 was chasing MiGs from the target when he was hit first by AAA, then a SAM. He ejected and was captured; subsequently released during *Operation Homecoming*.

Lieutenant (junior grade) William Metzger was flying an F-8C (BuNo. 147021) call sign "Page Boy 445" of VF-24 when he was hit by multiple rounds of AAA while flying at 1,500 feet. One round tore through the cockpit, breaking his right leg. He ejected and was captured; subsequently released during *Operation Homecoming*.

That afternoon an RA-5C Vigilante (BuNo. 150826) reconnaissance aircraft call sign "Flint River" of RVAH-13 *Bats* from *Kitty Hawk* was approaching the Hanoi TPP to obtain BDA photos. As the aircraft arrived over the city it was hit by intense and accurate AAA fire. With the Vigilante engulfed in flames and beginning to disintegrate, pilot Lieutenant Commander James Griffin and RAN Lieutenant Jack Walters ejected some 10 miles north of the capital city. They were immediately captured and taken to Hanoi Hilton, where both died several days later. The cause of their deaths is unknown, however, it is believed they sustained critical injuries in the course of ejecting.

It had indeed been a dark and costly day for American Naval Aviation. Six aircraft were lost, two from each carrier along with 10 men. It was an emotional gut punch for the sailors steaming on Yankee Station, one that would be felt for many years after. The day had been especially painful for the men of VA-35, who up to that point had not suffered a single casualty in the conflict. B/N Ken Van Lue, in a letter to his wife, wrote of the day "VA-35's luck finally has run out...it's getting to be a shitty war!" From that day forward 19 May 1967 was infamously referred to by naval aviators of Vietnam as "Black Friday".

Despite the heavy losses, the mission did have limited success. The strike group had managed to hit several targets which caused measurable damage, while *Bon Homme Richard* aircraft scored direct hits on the Hanoi TPP. The F-8 Crusaders that had been dispatched early in the mission ingress to intercept enemy aircraft shot down four VPAF MiG-17 fighters in the ensuing dogfights while losing two of their own. It was also good news that Red and Kelly were confirmed alive and without serious injury, with an accurate pinpoint of their location.

On standby for SAR was the Air Force's ARRS. The ARRS crews, stationed at numerous bases in Thailand, South Vietnam, and Laos, were ready to dispatch at a moment's notice. Typically when an aircraft went down and survivors were confirmed alive, a SAR effort was immediately launched, even if it was in an area with enemy activity. Once activated the standard response was two HH-3 Jolly Green Giant or HH-53 Super Jolly Green Giant helicopters accompanied by two A-1 Skyraider aircraft which would be dispatched from their bases or loiter areas. The Skyraiders would use their considerable firepower to clear the area of enemy activity and would stay on station for the duration of the rescue to provide additional assistance. Once the area was cleared of known threats one Jolly Green would come to a hover over the downed aircrew ("Jolly low"), and a jungle penetrator[6] lowered by winch. A Pararescueman (PJ) could be lowered in the penetrator to assist the survivor if he was injured or unable to do so himself. The second Jolly Green ("Jolly high") would hover at a slightly higher altitude nearby, providing suppressing fire and taking over in recovery, if necessary. For difficult rescues, additional air resources such as A-4 Skyhawks or F-4 Phantoms could be called in to help keep enemy ground forces at bay and provide RESCAP. In addition to ARRS, there were also clandestine

American *Lima* bases operating in eastern Laos as well as *Brightlight* and other SOG (Studies and Observations Group) assets. They were known to lend a hand in rescues, when they could discretely do so.

However, no rescue action was taken in the case of Red and Kelly that day. Citing "high SAM activity" over their location, command elements of the USAF 7th Air Force at Tan Son Nhut Air Base near Saigon prevented a rescue attempt from being launched. This was highly unusual as these units were known for their expertise and bravery in extracting downed aircrew from hostile areas. It was a bedrock of their existence, one they deservedly wore with great pride and distinction. There were numerous high-profile rescues during the conflict that justified this reputation. Many of these improbable rescues took place deep within North Vietnam and Laos, far into enemy territory.

The heat was nearly unbearable as the afternoon sun blazed overhead. Remembering the JEST training by Negrito tribesmen in the Philippines, Red tried to reach high ground where he could keep an eye out for approaching NVA forces. Being elevated would also increase the chances he would be able to contact and signal the rescue forces that would inevitably appear.

Though he had attempted to raise Kelly on the radio numerous times Red had received no answer. Their UHF radios were line-of-sight, and being separated by a high ridge meant their transmissions to each other were most likely blocked. Both could hear the radios of aircraft overhead, but not necessarily each other. He figured that once on high ground he would be able to contact Kelly, and possibly locate him. Off in the distance, he heard whistles, most likely from local militia or an NVA patrol searching for them. The whistles became fainter and fainter until he was no longer able to hear them. Drenched with perspiration and in considerable pain from back injuries, he began trying to climb to the top of the ridge that separated him from Kelly.

Later in the afternoon, after climbing about halfway up the ridge and completely exhausted, he called out Kelly's name. There was no answer. Another attempt on the radio also failed. Slumping against a tree trunk, he sprayed insect repellent on his exposed limbs and donned the mosquito net from his survival kit. His thoughts quickly turned to his wife and children; their tranquil home in Virginia Beach just a stone's throw from NAS Oceana. He imagined that soon a Navy staff car would be pulling up to the driveway. Red cringed at the thought of his family receiving the news he had been shot down and was missing. As for himself, there was no telling what was going to happen as his life was hanging in the balance. "God, why me?" he asked, while the full gravity of the situation began to materialize. Red, a man of deep Christian faith who attended church regularly, fruitlessly searched for an answer that was not yet there. With a throbbing pain in his back and distant, bygone images of domestic bliss filling his head, he dozed off.

He was awakened around 2200 by pouring rain. It was so heavy, coming in torrential downpours, it made an unmistakable roar as it pelted the heavy jungle canopy. The distinct sound of a propeller aircraft flying overhead momentarily broke through the crescendo, the strobe and beacon lights glowing softly above the low cloud cover. Thinking it could be an American aircraft he turned on his radio and tried to make contact. There was no response.

The rest of the night brought little rest for Red, who dozed in and out of sleep with every slight sound emanating from the jungle. He constantly glanced at his wristwatch, counting down the seconds until first light. What seemed like an hour was in reality only a few minutes.

Time seemed to stand still. Thirst and hunger pangs wracked his body. Hopeful thoughts of rescue turned, for the time being, into the reality of evasion and survival.

Back onboard *Enterprise* tensions in VA-35 were running high as Red and Kelly's squadron mates were becoming frustrated and angry at the lack of a rescue effort. XO Glenn Kollman and Nick Carpenter became very vocal and insistent to Commander Barie that something be done. It was these protests that resulted in plans for Carpenter and Slaasted to launch at first light the next morning in an A-6, accompanied by four F-4 escorts. Once confirmed the area was clear and Red and Kelly were ready for rescue they would again call ARRS and the Jolly Greens. Until then the two men were separated, injured, scared, and being hunted like animals in the thick jungle of the enemy's backyard.

Daybreak came at about 0530 on the morning of 20 May, a stagnant mist lingering throughout the jungle floor. A half-hour later Nick Carpenter and Richard Slaasted, their A-6A accompanied by F-4B escorts, flew low overhead. For Red and Kelly, it was a beautiful sight and sound. The planes were so close, it seemed they could almost reach out and touch them. In his after-action report, Nick recalled receiving two beepers as soon as they crested the valley where the two men were located.

Kelly was first to contact Nick, reiterating his badly broken leg. He also remarked that Red was "on the other side of the hill" and no enemy activity had been observed during the night. Flying over the area he got a visual on Kelly, who screamed "*Now! Now!*" on the radio when Nick was directly over him. Red then raised Nick on the radio and inquired about rescue. After a few minutes, Nick came back and replied that Jolly Green rescue helicopters from the USAF base in Nakhon Phanom, Thailand (known colloquially as "NKP," "naked fanny," and "no known place") were 45 minutes out. After several additional passes, the Intruder and Phantoms egressed from the area, having relayed all available information to ARRS.

The 45 minutes came and went, and two hours later there was still no sign of the Jollys. As late morning approached Red heard the first gunshot. He turned around to see two militiamen, clad in black with tire tread sandals, pointing their rifles at him. He had his .38 revolver in his hand and threw it to the side, which startled the men and caused one of them to fire again, closer to him than the first shot. Red immediately put his hands above his head. The two militia soon turned into a dozen, and he was a POW. Stripped of his few possessions and his arms tightly bound with rope, he was led away to an unknown fate.

Back on *Enterprise* Nick boarded an EA-3B Skywarrior of VQ-1 *World Watchers* call sign "Deep Sea One" and flew to Da Nang Air Base,[7] arriving at 1335 hours. Once there he checked in with the headquarters of the 480th TFS of the USAF. Somewhere up the chain of command, a decision had been made to try and bring at least one of the men out via a Fulton Extraction Kit. The kit consisted of a canister of helium, a large rubber balloon, 500-foot braided nylon tether line, and a torso harness. Once the harness was donned it was attached to the tethered balloon, which was inflated and rose high enough to be easily seen above the jungle canopy. A C-130 or similar aircraft with a v-shaped apparatus attached to the nose would then catch the tether, and the person in the harness snatched up and reeled into the aircraft. The system was originally designed for the CIA where it was known as the Surface-to-Air Recovery System (STARS).

At the headquarters of the 480th Nick met USAF Major Jim Craig, a burly Texan with a thick southern drawl who had experience driving an F-4 Phantom and dropping extraction and supply kits to U.S. special forces and CIA units operating along the Laos-North Vietnam border. Jim took an immediate liking to Nick and his devotion to bringing his squadron mates home. In a briefing that evening Nick pointed out to Jim the last known locations of Red and Kelly, and where the SAM had come from that brought them down. Nick would be flying in the back seat of Jim's F-4, traditionally occupied by a WSO. They would launch at first light.

On the early morning of 21 May Jim Craig and Nick were joined by another F-4 flying as wingman. As they crested the valley they picked up a solitary beeper. Thundering above the treetops in the mighty Phantom, Nick attempted to reach Red on the radio. There was only static until Kelly transmitted his call sign. Nick explained the extraction kit and the procedure for rescue. Kelly used his mirror to signal the F-4, and the kit was dropped about 100 yards down the hill from him. After flying over his position for another 45 minutes waiting for confirmation he recovered the Fulton Kit, they began receiving MiG warnings from a USAF EC-121 Warning Star surveillance aircraft call sign "Big Look". With heavy hearts, they headed back to Da Nang. Kelly must have felt helpless as he watched the Phantoms, spewing thick plumes of black jet exhaust, disappear into the horizon, leaving him alone once again.

Jim remembers that as he was flying over the valley where Red and Kelly were located the RHAW gear in his F-4 had remained quiet. However, when he climbed above the valley the RHAW began sounding off with SAM radar warnings. This indicated that low-flying aircraft in the area would ostensively not face any danger of SAMs. He also noted that during his flyovers he was not at any time targeted by AAA.

Commander Art Barie

Commander Glenn Kollman

Admiral Horace Epes

General William Momyer

Later that evening Nick again launched from Da Nang, this time aboard an Air Force C-130 Hercules aircraft, his fourth attempt to try and rescue Red and Kelly. They were joined by the same F-4s from the mission that morning, with Jim Craig and his WSO as flight lead. Once reaching the valley Kelly's beeper was picked up, and voice communication was established. Sadly, he had not retrieved the Fulton kit as hoped. The NVA captured it, and would have captured him in the process had he not retreated up the hill, barely escaping. Both Jim and Nick recall that Kelly sounded confused on the radio with some of his statements nonsensical. He spoke about seeing and meeting people in the jungle. When he was asked "is anyone with you?" he answered in the affirmative, and mentioned the name of a Norden civilian technician attached to the squadron. Either Kelly was attempting to talk in some impromptu code or his mental state was beginning to erode. This could have been a result of over two days in the unforgiving jungle with little food or water, a loss of blood, shock, delirium, or a combination of all. The disheartening fact that no rescue force had appeared must have also weighed heavily upon his mind. He must have also been thinking about Red, and whether he was alive, dead, or captured. Kelly helplessly watched as the three aircraft returned to Da Nang, with Jim and Nick dejected and filled with feelings of failure.

That night a conference call took place between Jim Craig, General William "Spike" Momyer, CO of the USAF 7th Air Force, and Admiral Horace Epes, CO of Navy Carrier Division One (COMCARDIV1). Jim told Momyer that he believed a helicopter rescue was possible and that enemy forces in Kelly's location were negligible. General Momyer, who had the final word on any rescue attempt in that area, stated a helicopter rescue mission was too dangerous and refused.

Jim recalls Epes exploding in rage at Momyer's refusal with a tirade of foul words only a Navy sailor dare conjure. Epes then stated if the Air Force would supply the helicopters the Navy would provide the crews and divert all fixed-wing assets needed for the rescue—the Air Force didn't need any of their people to get involved. Momyer stood his ground; there would be no rescue mission or aircraft from the Air Force.

The next morning, 22 May, a lone F-4B from *Enterprise* flew over the area and detected no locater beeper, voice communication, or enemy activity. Kelly was gone, MIA along with Red, who would not be acknowledged as a POW by the North Vietnamese until 1970. In a subsequent official report by VA-35 Skipper Commander Barie, he wrote:

> These men were down in good evasion/recovery country, but the location required deep penetration of rescue forces. An earlier decision and prosecution of rescue efforts may well have gotten at least one of these men out.

There was some degree of retribution for the men lost on 19 May. On 6 June an RA-5C Vigilante photoreconnaissance aircraft of RVAH-7 catapulted off *Enterprise*, heading to the Van Dien area southwest of central Hanoi that had become known as "SAM Alley". This was the same area where Red, Kelly, Commander Rich, and Lieutenant Commander Stark had been shot down three weeks prior. Pilot Lieutenant Commander Frank Hamrick and RAN Lieutenant John Capewell thundered over the area at supersonic speed, returning to *Enterprise* with a bounty of photographic evidence. PIs quickly spotted 10 SAMs and various components hidden beneath foliage, protected by

37mm/57mm and 85mm AAA. A plan was immediately put into motion to strike the SAM assembly and storage site the following day.

The weather on 7 June proved to be less than optimal for the strike, however so serious was the threat posed by these missiles that postponement was simply not an option. Commander James Shipman, CAG of CVW-9 led the strike in an A-4. He was joined by ten other aircraft consisting of A-4s from VA-113 and VA-56, F-4s from VF-92 and VF-96, and one A-6 from VA-35. To ensure destruction of the target a variety of weapons were carried. A-4s were outfitted with a combination of 2.75" rockets, deadly effective CBU-24 cluster bombs, and Mk.117 bombs. The F-4s were each carrying ten Mk.82 bombs in addition to two Sidewinder and two Sparrow air-to-air missiles to deal with any MiGs. The A-6 was carrying 22 Mk.82 bombs.

To assist in navigation around several cloud layers and a squall line of thunderstorms an E-2A Hawkeye of VAW-112 from *Enterprise* provided vectors through the weather. Utilizing a cloud-covered ridge to mask their approach from the south, fifteen miles from the target the section of VF-96 F-4s led by Lieutenant Commander Morton Winchester thundered ahead. They confirmed the missiles were still present, and commenced the attack. As the Phantoms rained down Mk.82 bombs the first-stage rocket boosters of several SAMs ignited, resulting in them wildly careening out of control and smashing into other SAMs and SA-2 components. It was already a site of sheer destruction when three rocket-equipped A-4s then made their runs, destroying another six missiles with direct hits. To ensure destruction A-4s and the lone A-6 followed with Mk.82, Mk.117, and CBU-24 bombs.

Three secondary explosions rocked out, sending fire and black smoke plumes into the sky. Shipman reconnoitered the area following the attack, remarking destruction of the target as "beautiful and great and satisfying".[8] Shortly after the site was destroyed a Vigilante from *Enterprise* made a BDA run over the target, piloted by RVAH-7 Skipper Commander Phil Ryan with RAN Lieutenant Jim Owen in the back seat. They were nearly shot down over the target, however, Ryan's skillful aviating got them back to the carrier safely. Once developed the photographs confirmed Shipman's observations.

Some 300 miles northwest of Yankee Station Red McDaniel was enduring relentless beatings, torture, and interrogation at the infamous Hoa Lo[9] prison (Hanoi Hilton). After several days an exhausted McDaniel was thrown into a cell with Bill Metzger, one of the F-8 pilots from *Bonnie Dick* who had been shot down the same day as he and Kelly. Red found him laying on a filthy dirt floor, barely conscious, with a gaping open wound in his left thigh that ran eighteen inches long and to the bone from a two-pound piece of AAA shrapnel that had penetrated the cockpit. He also had a broken right leg and deep lacerations on both arms. Metzger was hovering near death.

Despite having two crushed vertebrae in his lower back, Red picked up Metzger and placed him on a cot. He began nursing him back to health by tending to his wounds, feeding him, bathing him, and demanding to the guards that a physician see him immediately. Then, the answer to his question from that night alone in the jungle emerged. It was not "God, why me?" it was "God, why not me?" His purpose suddenly became clear, and throughout the following six years of imprisonment, he relied heavily upon and shared his religious faith with other prisoners.

VA-35 at the conclusion of the 1966-1967 cruise, after the loss of McDaniel and Patterson.

First Row: Townsend, Duff, Gordon, Berman, Sadowski. Second Row: Maso, Urbanek, Bremner, Fardy, Slaasted, Morgan, Hyde, Griffith, Malley, Dorn, Owen, Leonard, Turk, Gaynor, Bankson, Johnson, Borchers, Miles, Cable, Van Lue, Benjamin, Carpenter. Third Row: Barie, Kollman

In May 1969, after an unsuccessful escape by Air Force Captains Edwin Atterberry and John Dramesi, Red not only accepted the blame by his captors as a communications ringleader of the escape attempt, he also concocted a story about his own escape plan to prevent further punishment from falling on other prisoners. Despite days of torture that produced a compound fracture of his arm from being bound by rope tightly behind his back and weeks of solitary confinement, he never revealed a single name or thread of information. Atterberry was tortured to death as punishment, dying on 18 May. Dramesi barely survived the brutal torture and was repatriated during *Operation Homecoming.* After being released from solitary confinement back into the general prisoner population, sensing discouragement amongst his fellow prisoners, Red led a mass protest for religious services, causing a headache for his captors. He was eventually allowed to lead Sunday church services, becoming a makeshift chaplain and counselor to the other inmates, many of them terribly ill. For his heroism during internment, Eugene McDaniel was awarded the Navy Cross.

Following his release and a prolonged period of rehabilitation at Portsmouth Naval Hospital Red returned to active duty. He eventually attained the rank of captain and CO of the aircraft carrier USS *Lexington* (CVT-16). His final assignment in the Navy was liaison to the United States House of Representatives. He retired in 1982 after 27 years of service and 81 combat missions over SEA. Following his Navy career he headed the American Defense Institute, a think-tank focused on critical issues of national security. He regularly gives public speeches on his experiences in the Navy, and the years he spent as a POW in Hanoi Hilton. Amazingly Red holds no resentment, bitterness, or hatred towards his captors stating "If I did, it would have eaten me alive many, many years ago".

Colonel Jim Craig retired from the Air Force in 1982 with 150 combat missions to his credit, including daring Walleye bombing missions against heavily defended bridges north of the DMZ. He describes the events of 21 May 1967:

> That was the darkest day of my deployment, perhaps of my entire Air Force career. Kelly was right below us, injured and scared. There we were, in a multi-million dollar fighter, just a few hundred feet away, absolutely powerless to rescue him. My disappointment has not abated one bit in the decades since.

When *Operation Homecoming* commenced in February of 1973 for some it brought jubilance, others closure, but for hundreds of other families, it brought neither. Kelly's parents eventually approved a change in his status in 1974 from MIA to Presumptive Finding of Death (PFD). They begrudgingly accepted the Navy's determination that Kelly died in the jungles of North Vietnam, most likely at the hands of local militia. Following his change in status Kelly was posthumously promoted to lieutenant commander.

Upon his arrival at Hanoi Hilton, Red repeatedly inquired about Kelly. In late 1967 a guard informed him Kelly was alive and recovering from injuries. It made sense, as Red knew Kelly had been injured from the radio conversations with Nick Carpenter. In a follow-up inquiry, he was told Kelly was fully recovered. However, further inquiries yielded inconsistent information on his friend, which left him with the conclusion that the guards were being sadistic and lying. After so many years of his ordeal as a POW he begrudgingly accepted the conclusion that Kelly was dead.

When official U.S. inquiries were made into Kelly's fate, the North Vietnamese response was total silence. Despite this dodge, USAF RF-4C PSO 1st Lieutenant Ronald Mastin,[10] who became a POW after being shot down on 16 January 1967 reported that during his captivity in Hanoi, he saw what he believed to be a photo of Kelly's ID card in a North Vietnamese newspaper. Despite this report, U.S. investigators have not located a newspaper of that period with Kelly's picture. Additionally, the file of repatriated USAF F-105 Thunderchief pilot Colonel Dewey Smith mentions Kelly's name being on a POW interrogation questionnaire. There was also an alleged sighting of Kelly's name being scratched into the wall of a North Vietnam prison cell. However, as with the purported picture of Kelly in a North Vietnamese newspaper, this claim has not been substantiated.

In 1985, after eighteen years of silence, the Vietnamese suddenly produced Kelly's military identification along with his Geneva Convention card. When pressed for details the Vietnamese proffered several accounts of Kelly's fate. From November 1987 to January 1990 the American government received and cited multiple North Vietnamese reports stating Kelly was killed during an exchange of gunfire with North Vietnamese militia and promptly buried in the jungle. These reports were written by Vietnamese authorities more than a decade prior but were only made available to the American government in the late 1980s. There is credence to these initial accounts as Kelly had previously stated that his greatest fear was capture, therefore he was likely to resist any attempts to take him. However in February 1991, the Vietnamese provided several alleged eyewitnesses who stated there was no firefight; Kelly was found hiding in the jungle, and with no resistance, he was killed by a single shot to the chest, his body buried where it fell. These accounts directly contradict the earlier official written reports of resistance to capture and subsequent gunfight.

The eyewitnesses also led U.S. investigators to Kelly's alleged burial site, but when thoroughly excavated by Joint Task Force-Full Accounting (JTF-FA) team members they found no remains and no disturbance to the soil strata, indicating no burial had ever taken place in that area. No bodily remains of Kelly or wreckage of Ray Gun 502 have ever been located. The ultimate truth of what happened to him, and so many other Americans missing in SEA, wherever it may lie, is likely to never be known. Classmate Lee Cargill still thinks of Kelly after all these years:

> Life goes on but we do not forget friends like Kelly. There was a student nurse in Philadelphia that he was in love with. I cannot remember her name but do remember she broke off the relationship with him when I believe she became interested in a fellow who was a medical student. Kelly was heartbroken. I wonder to this day if she ever knew of his loss. I thank God for the privilege of knowing Kelly. The world is a better place for him having traveled a short time in our midst.

Jim Ring adds:

> How sad, all in the Class of 1963 have missed him at all our reunions over the years. It would have been great to see him excel in whatever he might have done in life because I know it would have been great.

Squadron mate Bill Gaynor reflects solemnly on Kelly's loss:

> I was not on that mission to Van Dien but was certainly
> affected by the aftermath of the sortie and the loss of
> Kelly. I spent quite a bit of time with him, both in the
> squadron and on leave. He most definitely was a person
> who enjoyed and lived life to its fullest, and I consider it a
> privilege to have known him during his short time here.
> Kelly will always be in my memories.

In June 1985 surviving members of USNA Class of 1963 gathered at
the Vietnam Veterans Memorial in Washington, DC. They came
together to honor Kelly and other classmates who never returned
home[11] from Southeast Asia. Classmate Captain Michael Cronin,[12]
USN (ret.) gave a solemn speech where he stated:

> Now after all these years, it may be easier to understand
> what they lost. Look around, they lost all that we have
> had over these many years. They lost the chance to have
> families, careers, dreams, hopes, victories, failures,
> disappointments, and even love.

Following that first deployment, on 14 July 1967 at a ceremony at
NAS Oceana, Commander Glenn Kollman relieved Commander
Arthur Barie, becoming skipper. Commander Herman Turk assumed
the role of XO. In September 1967 VA-35 flew to San Diego for pre-
deployment work-ups aboard *Enterprise*. While there President Johnson,
Secretary of Defense McNamara, and CNO Admiral Thomas Moorer
came aboard. As part of their visit, Skipper Kollman hosted them in the
ready room and gave a briefing on the events of 19 May.

He presented a complete rundown of the mission, including the loss of McDaniel and Patterson. Kollman and others in the squadron at the meeting hoped at least one of the three men would ask the question: Was the result worth the lives lost? They did not.

The New Year had barely rung in when *Enterprise* departed NAS Alameda on 3 January 1968. After stops in Pearl Harbor and the Philippines, the customary two-week transit to the Tonkin Gulf was interrupted by the 23 January seizure of the Navy ship USS *Pueblo* (AGER-2) by North Korea.

On 31 January North Vietnam launched the Tet Offensive, a massive operation that caught the United States and South Vietnam off-guard. While extensive Viet Cong guerrilla attacks were carried out throughout the south, the DMZ area around the Marine Corps forward fire base of Khe Sanh came under continuous siege by the NVA. Standing by for a possible confrontation with North Korea, *Enterprise* held west of Sasebo, Japan for nearly a month. After a short stop in Subic Bay, the air wing began flying combat sorties from Yankee Station on 22 February, three weeks after the start of the Tet Offensive.

On 28 February pilot Lieutenant Commander Henry Coons and B/N Lieutenant Thomas Stegman call sign "Ray Gun 512" were flying a night strike against coastal defenses at Do Son in North Vietnam. The men made a routine radio call stating they were 14 minutes from the target and about to begin their run-in. That was the last communication from them. After a massive land and sea search wreckage was spotted 20 miles off the coast, east of Thanh Hoa. Among the items floating in the water were parts of the empennage, peppered with flak damage. It is assumed the Intruder (BuNo. 152938) took a direct hit. Both men were ultimately declared KIA.

Two days later, on the night of 1 March (1968 was a leap year), three VA-35 Intruders were launched from *Enterprise* against targets in North Vietnam. Pilot Lieutenant Commander Edwin Scheurich and B/N Lieutenant (junior grade) Richard Lannom call sign "Ray Gun 504" were targeting an army barracks in the coastal city of Cam Pha, northeast of Haiphong. The aircraft (BuNo. 152944) vanished without a trace. In 2017 remains were discovered on Tra Ban Island, off the coast of Cam Pha. The remains were later confirmed to be those of Lannom. From the location of the remains, it is theorized that in the darkness of night, the aircraft failed to clear a high peak. However, enemy activity cannot be excluded as a factor.

Less than two weeks later, on 12 March, the squadron lost Skipper Kollman along with B/N Lieutenant John Griffith. Their Intruder (BuNo. 152943) had just launched from *Enterprise* in poor weather when the aircraft pitched up sharply, nosed over, and crashed into the sea. The aircraft quickly sank, with Kollman and Griffith still strapped into their seats. Their remains were not recovered. Though numerous causes have been speculated such as a dislodged VDI or premature retraction of the flaps and subsequent loss of lift, in all likelihood it was a faulty leading-edge slat that had been problematic for the aircraft in previous flights.

The loss of Skipper Kollman was a terrible blow to the *Black Panthers*. He was admired and respected by the squadron, and known to lead the most difficult and dangerous missions, never asking his men to do something he hadn't. In an interview with *Naval Aviation News* published posthumously in August 1968 Griffith wrote of Kollman:

> Commander Kollman described himself as a "pacifist by nature" but added that he never sought an excuse to justify his presence in combat. "My reasons are simple,"

he said, "they are my moral obligation and commitment to our nation. I identify myself with the obligations of the United States to honor its commitments with foreign nations all over the world. To me, my duty here is as natural and normal as accepting my responsibilities for caring for my wife and children. It is my job, and I'm going to do it. I wouldn't ask someone else to do my job".

Griffith was a highly talented man. In addition to being an A-6 B/N, he was also an architect, artist, and journalist. Just sixteen days before their loss, on 24 February, Kollman and Griffith had led a daring night strike against a port facility in Hanoi, the heart of North Vietnam. Despite poor weather and a barrage of deadly and accurate AAA fire and SAMs, the men led the Intruders to their target and inflicted devastating damage without suffering a single casualty.

The day after Kollman and Griffith were lost CINCPAC Admiral John Hyland came aboard *Enterprise*. The primary reason for his visit was to preside over an awards ceremony, where the Navy Cross was to be awarded to both Kollman and Griffith for the 24 February mission. Instead, Admiral Hyland expressed his condolences and awarded the medals posthumously. After the loss of Kollman, it was Commander Herman Turk who took the helm as skipper, with Commander John Frick replacing him as XO.

Five days after Kollman and Griffith were lost, on 17 March, a flight of four Intruders launched on a low-level night mission to targets within North Vietnam. Pilot Lieutenant Commander Arthur "Ned" Shuman and B/N Lieutenant Commander Dale Doss call sign "Ray Gun 510" were assigned a railway yard at Khe Nu, ten miles north of Hanoi. The men were flying north of the capital city en route to the target when their Intruder (BuNo. 152940) was hit by large-caliber AAA. The windscreen of their aircraft was shattered and, along with other damage, forced them to eject. Both men were immediately captured and released five years later during *Operation Homecoming*. Shuman was on his 17th mission when he was shot down.

Nearly two months later, on 13 May, pilot Lieutenant Bruce Bremner and B/N Lieutenant Jack "Dancin' Jack" Fardy call sign "Ray Gun 510" were flying a low-level night strike on Vinh Airfield. Their ordnance was 18 Mk.36 DST magnetic-fused bombs. After dropping the weapons they were hit by 57mm AAA in the port wing, which caught fire. The men somehow nursed the burning Intruder (BuNo. 152951) back to *Enterprise*. As they were descending an explosion rocked the aircraft, forcing their ejection. Both men were successfully rescued without injury by the carrier's plane guard helicopter.

On the previous VA-35 cruise 1966-1967, Bremner had permitted B/N Fardy to reach over and get a feel for the control stick. This was not unusual in the A-6 as there always existed a chance of pilot incapacitation where the B/N would need to temporarily "fly" the airplane. Fardy was too aggressive and put the aircraft in a dive that Bremner struggled to recover from. With the Intruder descending steeply, he ordered Fardy to eject. Shortly after that, he was able to regain control and landed aboard *Enterprise*. Fardy was rescued from the water without injury. On 19 April 1971, Bremner experienced a second ejection when he and his B/N were forced to punch out of their Intruder (BuNo. 155270) due to a loose stabilizer while operating from NAS Pax River. Both men escaped unharmed.

The squadron's last loss of the cruise occurred on 24 June, less than a month before the end of the deployment. Pilot Lieutenant Nick Carpenter and B/N Lieutenant (junior grade) Joseph Mobley, call sign "Ray Gun 503," launched on a low-level night mission to lay maritime mines in the Song Ca River, some 5 miles southwest of Vinh. As they were approaching the target waterway at an altitude of 250 feet and traveling at a speed of 420 knots (483mph) a AAA shell burst close to their Intruder (BuNo. 152949), spraying shrapnel into the cockpit. Carpenter was severely injured, perhaps mortally, and the aircraft began rolling uncontrollably. Mobley ejected and was captured. He was released during *Operation Homecoming*, eventually reaching the rank of admiral before retiring in 2001. Carpenter remained MIA until his remains were repatriated in January 1989 and positively identified in April 1991. Posthumously promoted to lieutenant commander, he is interred at Arlington National Cemetery. Like Skipper Kollman, Carpenter's loss deeply affected the squadron. He was also widely admired and respected, recognized as a highly talented pilot and good friend. He had shown great care and concern for his squadron mates when he pressed for, and volunteered to take part in, the rescue efforts of Red and Kelly the previous year. He is remembered often by his family and squadron mates, who treasure his memory. His cousin Meredith reiterates the profound impact his loss had on their family:

> Nick's Father never recovered from him being MIA, not knowing for so many years what happened to his son. Sadly, his Father died before he was finally brought home in 1989. We all miss Nick dearly. He was so much fun, full of life. More than a half-century later we still have a hard time accepting he is gone.

69

Men in the air wing recall Nick's integral part in spearheading a rescue effort of Red and Kelly, trying so valiantly to bring them back home. Paul Daley pointedly remarks:

> What Nick Carpenter did to try and rescue Red and Kelly was above and beyond the call of duty. He repeatedly and selflessly put his own life on the line trying to get those men out. He is truly a hero.

The 1968 deployment was a difficult one for VA-35, suffering seven fatalities in just four months of frenzied action. The *Black Panthers* would deploy two more times to Yankee Station, aboard USS *Coral Sea* (CVA-43) October 1969-June 1970[13] and USS *America* (CVA-66) July 1972-March 1973.[14] The latter cruise saw the squadron taking part in *Linebacker* operations that forced the North Vietnamese to the negotiating table in Paris, resulting in the 27 January 1973 ceasefire.

---

[1] Within the air wing IOIC was sometimes referred to as "101 Clowns".

[2] Topographical formations as a result of the erosion of soluble minerals, creating caverns, crevices and caves. Karst formations can tower many hundreds of feet above ground level and are often covered in thick vegetation.

[3] Document carried by military airman, printed in local languages, offering civilians monetary and other rewards for assistance, protection and safe return.

[4] When Fan Song radar acquired a target the crew would hear slow, siren-like warble tones in their headsets ("singer low"). Once the Fan Song radar pulse repetition frequency changed from acquisition to guidance the warble tones increased to a faster tempo ("singer high"). This meant one or more missiles had been launched and were being guided to their targets.

5 Junior officers in VA-113 had previously been accused of using the radio excessively during missions, thus blocking critical communications. A junior officer in the squadron connected a tape cassette recorder to the A-4 radio system to gather evidence it was senior officers using the radio excessively. It was by happenstance he captured the turmoil.

6 The jungle penetrator was shaped like a bullet, approximately four feet long and rounded on the bottom, allowing for easy penetration of thick jungle canopies. Three spade-shaped paddles folded down from the body, making it resemble a large grappling hook. The paddles could be used as seats, or to grasp on to. Heavy straps attached to the body of the penetrator gave a way for survivors/rescuers to safely secure themselves during insertion/extraction.

7 Da Nang Air Base was used by every branch of the U.S. military as well as the VNAF and commercial airline carriers. At the height of the conflict in 1967-1968 it was one of the busiest airports in the world.

8 *Enterprise* 1967 Command History

9 Pronounced *"Wah-lo"* the prison was built by French colonists in the late nineteenth century to inter political prisoners.

10 See Chapter IV endnote 51.

11 The first combat loss in Vietnam for USNA Class of 1963 was Lieutenant (junior grade) Carl Doughtie. Piloting an A-1 Skyraider (BuNo. 137521) of VA-25 *Fist of the Fleet* from USS *Midway* (CVA-41), he was killed on 10 June 1965 while bombing a power plant southwest of Thanh Hoa. Unable to pull out of a dive on the target, his aircraft impacted the ground and exploded. His remains were repatriated on 30 September 1997, and interred at Arlington National Cemetery on 25 February 1999.

12 See Chapter IV endnote 53.

13 The only loss VA-35 experienced on this deployment occurred on 26 December 1969 when pilot Lieutenant (junior grade) Dustin Trowbridge and B/N Lieutenant (junior grade) Walter Kosky were conducting tanker operations when their A-6A (BuNo. 152891) crashed into the Tonkin Gulf for unknown reasons. Both men were killed.

---

[14] VA-35 lost two aircraft and two men on the squadron's last deployment of the conflict. On the night of 16 December 1972 pilot and squadron Skipper Commander Verne Donnelly along with B/N Lieutenant Commander Kenneth Buell were conducting a night recce mission near Hanoi when their A-6A (BuNo. 157208) call sign "Ray Gun 504" was hit by AAA and exploded, with both men killed. Donnelly had taken over command of the squadron just seven months prior. His remains were repatriated and identified in February 1991.

On 24 January 1973 pilot Lieutenant C.M. Graf and B/N Lieutenant S.H. Hatfield were on a CAS mission when their A-6A call sign "Ray Gun 507" (BuNo. 157007) took ground fire six miles north of Quang Tri. They were feet wet on their way back to *America* when they were forced to eject. Both men were picked up by a Navy helicopter.

# Chapter II

# The Legend of Lucky 7

October 13, 1962 was a quiet Saturday evening for North American Aviation A-5 program manager John Fosness, who was relaxing in the comfort of the Columbus, Ohio home he shared with his wife and two young daughters. The quiet was suddenly interrupted by the loud ringing of the telephone.

On the other end of the line was a high-ranking official calling from the Pentagon who was curt, and very much to the point. Though posed as a question in reality there was little room but for one answer. At the North American factory at Port Columbus Airport sat two prototype aircraft in the process of conversion from A-5 nuclear bombers to RA-5 photoreconnaissance aircraft. Still loaded with development equipment and not yet combat operational, the Pentagon official inquired whether the two aircraft could quickly be made ready with electronic warfare equipment and basic reconnaissance cameras. John told the man yes it was possible. The official then gave him and his team only 24 hours to deliver the two completed aircraft to NAS Key West, Florida.

Frantically calling members of his team, John had little luck finding anybody at home on a Saturday night. Babysitters dutifully relayed the location of parents out for an evening on the town, while several dinner parties were interrupted by the same ringing of the phone that had earlier shattered John's quiet evening. By midnight he had managed to gather about a hundred members of his team at the North American factory to begin their monumental task.[15]

As they worked into the morning, some 2,000 miles to the west at Edwards AFB in the California desert, Air Force Major Richard Heyser lifted into the sky in a top-secret U-2 "Dragon Lady" (SerialNo. 66675) reconnaissance aircraft. By afternoon Eastern Time on Sunday 14 October 1962 American intelligence specialists were looking at photos of Soviet nuclear missiles stationed in Cuba. The crisis had begun.

Two weeks later Khrushchev and the Soviets blinked, and the crisis ended with the RA-5s never leaving NAS Key West. The two prototype aircraft were flown back to Columbus to complete their testing. Though temporarily deprived of the opportunity to showcase its reconnaissance talents over Cuba,[16] in a short time the RA-5 would rise to meet its destiny, courtesy of a highly skilled group of naval aviators. Together these men, and the aircraft, would become legends of Naval Aviation. They would soon make history in a war-torn region of the world known as Southeast Asia.

The North American Aviation RA-5C Vigilante reconnaissance aircraft, or "Vigi" as it was affectionately known, was unique to Naval Aviation during the Vietnam conflict in several ways. With an official top speed of Mach 2, it was the fastest aircraft to serve aboard an aircraft carrier.[17] A wing area of 701 square feet, a length of over 76 feet, and a maximum gross takeoff weight of around 80,000 pounds also made the Vigi one of the two largest and heaviest aircraft deployed on a carrier, the other being the A-3 Skywarrior.

Due to the large size and weight, as well as unique control characteristics, it was a challenging aircraft to fly. There were no ailerons on the Vigilante, instead spoilers and vents on the wings provided roll control and also acted as speed brakes. The spoilers and elevator, which were two solid slabs in a stabilator configuration, controlled pitch and were adjusted separately for roll trim. The rudder was also one slab rather than a hinged vertical stabilizer design. These gave the Vigilante flight control characteristics, unlike any other aircraft in the fleet. The unique control systems along with the high angle of attack and fast approach speed required for landing tended to challenge even the most seasoned of naval aviators. With the unwieldy size, a nose gear that was a full nine feet aft of the pilot, and the unique roar of its two Pratt & Whitney J79 engines when it came aboard or departed the carrier, jokingly compared to the sound of a pachyderm in heat, it was given the endearing nickname "Elephant". Indeed, the process of positioning Vigilantes on the carrier flight deck was known as the "Waltz of the Elephants." Despite this cheeky moniker, the RA-5C may very well be the most beautiful aircraft to ever grace the deck of an aircraft carrier.

Originally envisioned in the early 1950s as the A-5, a carrier-deployed supersonic nuclear bomber, the success of the Navy's submarine-based Polaris missile had negated its intended role. However, the intensifying Cold War and escalating conflict in Vietnam repurposed the Vigilante as a reconnaissance aircraft, fortunately saving it from the boneyard of Cold War aviation history.

The Vigilante carried a crew of two—a front-seat pilot and a rear-seat RAN. Flying an RA-5C in the Vietnam theatre took a great amount of skill and concentrated effort on the part of the crew. The pilot had to constantly maneuver, and "jink" at least 45 degrees every 15 seconds along a pre-determined route to the target to avoid ever-present AAA and SA-2 SAM missiles. If over a large city or high-value target RHAW threat detectors would give the pilot and RAN continuous warnings of being tracked and locked onto by these deadly weapons, with the RAN continuously feeding the pilot vital information through the ICS. It only took a momentary lapse in concentration from either man to imperil the mission and their lives.

The RAN in the back seat sat behind a console that was a myriad of warning lights, dials, switches, knobs, and analog counters, as well as various scopes, screens, and an optical viewfinder. He was responsible for not only navigation, radar, and electronic warfare, but also the paramount task of proper configuration and operation of the sophisticated photoreconnaissance equipment. He had to monitor film usage, and camera exposure and adjust the critically important Image Motion Compensation (IMC). The IMC was viewed through the optical viewfinder found on the belly of the fuselage, which showed the RAN the ground with the image superimposed with moving lines of light. The RAN had to constantly adjust the lines to match the aircraft's trajectory to obtain the best images. This took a great amount of concentration, and attention to detail.

While Vigi pilots praised the excellent visibility from the front seat, the backseat RAN worked his dizzying array of equipment in near-total darkness. When designed as a nuclear bomber the B/N sat in a windowless canopy meant to protect him from the blinding flash of a nuclear detonation. This would allow the B/N to fly the airplane should the pilot become temporarily blinded or incapacitated. The windowless RAN canopy was also meant to shield sensitive avionics from EMPs associated with nuclear weapon detonations. When redesigned for reconnaissance two small windows were added to both sides of the RAN canopy, with retractable blinds that covered them during missions, thus improving the visibility of the numerous instruments.

Many features of the RA-5C Vigilante were revolutionary and state-of-the-art for the time, having initially been designed to operate in and withstand, the rigors of nuclear war. At the heart of the avionics package was the North American Autonetics AN/ASB-12. This package featured a multi-mode Radar-Equipped Inertial Navigation System (REINS) originally developed by North American for the SM-64 Navaho ICBM.

The AN/ASB-12 utilized one of the first HUDs and was capable of low-level, terrain-following flight using only radar guidance. A close-circuit television system using a video camera attached to the belly of the fuselage projected the image to both the pilot and RAN. The advanced avionics were run by the Versatile Digital Analyzer (VERDAN), which was located under the seat of the pilot. It was one of the first solid-state computer systems ever fitted to an aircraft. The Vigilante was also one of the first aircraft equipped with a fly-by-wire system versus traditional cable and pulley controls.

The airframe design of the Vigilante was equally impressive and optimized for supersonic speed. High-stress areas were made from pure titanium. The wing skins were single pieces of metal milled from aluminum-lithium alloy. The pilot's windscreen was made from a single sheet of stretched acrylic, designed to be impervious to bird strikes. Gold plating in the engine bays helped reflect heat away from the power plants. Engine inlets were optimized for peak supersonic efficiency, and adjustable for different Mach speeds, while leading-edge droops critically improved slow-flight performance.

The versatile Vigilante performed three crucial reconnaissance roles in the Vietnam conflict—optical, electromagnetic and electronic. As part of the process of transforming the aircraft from bomber to reconnaissance, an aerodynamic canoe-shaped fairing was attached to the centerline belly of the fuselage. The canoe could hold numerous pieces of reconnaissance equipment, with the configuration depending on the mission. Two different spinning-prism panoramic cameras could be pivoted to permit photos either straight down or from side to side. When sweeping horizon-to-horizon the panoramic cameras were synchronized to the speed of the aircraft. The low-altitude KA-57A camera used a 3" focal length for 3,000 feet while the high-altitude KA-58A camera used an 18" focal length for 30,000 feet. The images from the 18" camera were so detailed it could capture a baseball-sized object from 20,000 feet and was also used in the top-secret U-2 spy plane. The film negatives from these cameras were massive, with thousands captured every mission. The Vigi used so much film that during peak years of the conflict, the U.S. Navy was one of Kodak's largest customers.

In addition to the panoramic cameras, there was also a KA-50A, KA-51A, or KA-62A vertical camera with a 36" focal length that shot straight down. Along with panoramic and vertical cameras, there was a forward camera that fired during the entire mission, documenting the route of flight. There were also two KA-51A/B oblique cameras, one on each side, that could be set to a variety of angles. Flasher pods could be mounted in wing stations, providing high-intensity flashes of light during nighttime or low-light photoreconnaissance missions. However, flasher pods were rarely used. When mounted on a wing station they severely restricted the speed of the Vigilante. The flashes also acted as a beacon for AAA. An IR sensor package made available in 1968 largely negated the need for flasher pods.

SA-2 SAM detonation as photographed by Vigilante oblique camera.

Good photography is an art form, and this is especially true of RA-5C photoreconnaissance missions in Vietnam. Simply flying over or near their objectives with the cameras firing was not always sufficient. Both pilot and RAN had to know how to get the best pictures in any given environment, which was a skill unto itself. Considerations such as speed, altitude, bank and pitch angle, film exposure, terrain shadowing, and the position of the sun all factored in obtaining the best intelligence photos. All of these variables had to be taken into account, regardless of how deep inside enemy territory they were, or threats present such as AAA and SAMs. In fact, photographs of these threats were often the objective of their flights.

With the "stores train" linear bomb bay no longer being used for munitions it instead held additional fuel and, if the mission so required, an AN/ALQ-61 PECM unit for ELINT. Though not visible during a cursory glance, the Vigilante was lined with antennas, from the pitot tube in the nose to panels in the skin of the fuselage and leading edges of the wings. These antennas were connected to RHAW receivers which detected specific radar bands associated with North Vietnamese (Soviet) weapon systems such as Spoon Rest and Fan Song used by the SA-2. RHAW receivers would alert the crew with audible and visual warnings, with the RAN using the AN/APR unit to determine bearing and distance to origin. When equipped with the AN/ALQ-61 the pulse-repetition frequency and bandwidth of the detected radar were recorded on a magnetic tape. The Vigilante also carried a package of ECM equipment that was operated by the RAN, which used several methods to "spoof" or otherwise disrupt AAA and SAM radar as well as block radio command and control signals transmitted to missiles.

If the mission so required an AN/APD-7 SLAR occupied the rear of the canoe. The SLAR could be utilized to make detailed maps which were used, amongst other purposes, to plan low-level A-6 Intruder missions. SLAR did not require daylight, therefore these missions could be flown at night when there was less chance of enemy ground fire.

To process this huge amount of data the squadron intelligence department worked out of several connected compartments in IOIC, which was located just below the flight deck, managed and supervised by ship's company.[18] IOIC was where the ship and squadron staff processed intelligence, supported aircrew mission planning and briefed/debriefed flight crews. It was also where sensor data was reviewed and analyzed, used to update the order of battle and target databases as well as prepare intelligence and operations debriefs which were produced after every cycle into formatted OPREPs. Though each squadron had a ready room, the centralized resources of IOIC is where the majority of pre and post-mission activity took place.

The OPREPs would be disseminated throughout various command staff elements, from the task force flag admiral to individual squadrons. They were also forwarded to CINCPAC, CINCPACFLT, and the Headquarters of the Pacific Air Forces (PACAF). They were also sent to the Pentagon and senior personnel in Washington, DC. Typical heavy reconnaissance squadron intelligence staff consisted of four AIOs and a dozen enlisted Photographers Mates (PH), Photo Technicians (PT), and Photo Interpreters (PI). Select AIOs were specialists in certain intelligence aspects such as PECM, imagery, targeting missions, maintaining the target database, ensuring compliance with the ROEs, and defining BDA mission requirements.

After a Vigilante returned to the carrier from a reconnaissance flight squadron flight deck personnel would remove the exposed, undeveloped film magazines and PECM magnetic tapes, taking them to IOIC. For photo images, a Kodak EH-38 processing machine would quickly develop the film, with the resulting positive film prints arranged on a Bausch & Lomb light table for analysis. In addition, a 10X microscope-like magnifying device was also used for detailed analysis. IOIC was also equipped with two Stereo Comparison Viewers (SCV), which could take two pictures and overlap them in such a way as to create a three-dimensional image when viewed through a stereoscope.

The Digital Data System (DDS) used data from the INS to tag images from the cameras and PECM from magnetic tape with the date and time, aircraft heading and altitude, and geographical coordinates of where the image was taken or signal received. This metadata was part of the Coded Matrix Block (CMB) and was crucial in expeditiously analyzing the voluminous amounts of data gathered on each mission. A Vigilante reconnaissance squadron could generate as much as 10,000 feet of film and magnetic tape per day. To assist in sorting and assembling the mountainous trove of intelligence data IOIC had an IBM punch-card computer that would read and process the CMB metadata from photos and PECM, and logically correlate the streams of data together. The resulting analysis drew an "Electronic Order of Battle," providing a highly accurate map of locations and types of enemy AAA and SAM radar.

The reconnaissance missions of the unarmed Vigilante were among the most dangerous flown by combat aircraft during the conflict, particularly pre and post-strike BDA flights, which were the majority of RA-5C missions in theatre. For these sorties, the Vigi would fly just 100-200 feet above the ground to an IP, where the pilot would accelerate to Mach 1.2 (913 MPH) and climb steeply to 3,500 feet above ground level for their BDA runs. Though the aircraft was designed for, and perfectly capable of high-altitude reconnaissance, reverberations of Francis Gary Powers being shot down in a U-2 over the Soviet Union lingered in the Pentagon. They did not want the sophisticated aircraft to fly at higher altitudes, which would make it susceptible to the SA-2 SAM, the same that had brought down Powers in 1960. Instead, it flew low, where the dangers of a SAM were less but the AAA threat was much greater. Fear of losing a Vigilante was also why, when first deployed to Vietnam in 1964, RA-5C squadrons were restricted to operations south of the DMZ, flying from carriers on Dixie Station.

This sensitivity to losing an RA-5C in combat was demonstrated when the first Vigilante (BuNo. 149306) lost in the conflict went down in South Vietnam on 9 December 1964. Pilot Lieutenant Commander Donald Beard and RAN Lieutenant (junior grade) Brian Cronin of RVAH-5 *Savage Sons* had catapulted from USS *Ranger* (CVA-61) on Dixie Station for a nighttime flasher pod reconnaissance mission of Highway 1 when the aircraft was lost to undetermined causes.

As the cause was unknown the Navy was anxious to examine the wreckage and retrieve several pieces of sensitive equipment. Vigi pilot Lieutenant Jim "Pirate" Pirotte, who knew the aircraft intimately and was also the maintenance officer of RVAH-5, was selected for the task. A SOG team was first sent in, consisting of an American special forces operator accompanied by a team of Montagnards, to find the wreckage. It took three weeks to locate the aircraft remains which were in a remote, mountainous area of South Vietnam infiltrated by VC. Once the perimeter was secure Jim was brought in via helicopter and lowered by winch into the thick jungle. He later recalled the story to his son Jay:

> Dad said he took a carton of cigarettes to the special forces operator and the guy looked like Rambo, with bandoliers of ammunition crisscrossed across his chest and numerous knives strapped to his body. The man remarked that catapulting off a carrier and flying combat missions over Vietnam in an unarmed aircraft would scare him to death. Dad remembered that as the man was saying this he was thinking to himself that in 45 minutes he will have a hot shower and a fresh-cooked meal in the officer's wardroom, while it will take this guy three weeks of fighting back to base camp just for K Rations.

Vigilante missions were common over heavily defended cities such as the capital Hanoi and the port city of Haiphong, as well as fortified enemy targets throughout the country. While pre-strike BDA flights often caught the North Vietnamese off-guard, post-strike BDA flights were a different story. The North Vietnamese anticipated these flights and were often prepared with AAA. With the RA-5C routinely flying unarmed into the belly of the beast, there were very few "milk runs" for Vigilante crews.

A recipient of numerous speed and altitude records,[19] the Vigi was able to outrun most threats. Despite this, they always flew with an F-4 fighter escort. However many Phantom pilots were not comfortable flying so low to the ground for the run-in to the IP for BDA missions. In addition, the Phantom had a 600-knot speed restriction if equipped with external fuel and munitions, while the RA-5C was a clean airframe that flew supersonic for their BDA runs. Some F-4 pilots ignored the speed restriction and tried to keep up, while others knew where the Vigi was headed and went to a pre-determined rendezvous point to meet after the mission. This arrangement was often made during pre-flight briefing in IOIC, in conjunction with the F-4 pilot and their back-seat RIO. Consequently, the aircraft and crew were often vulnerable and unprotected during the most dangerous phases of missions deep within enemy territory.

The RA-5C proved itself a highly effective reconnaissance platform in Vietnam soon after its arrival in August of 1964 aboard *Ranger*. By the end of that year, all existing A-5 bomber aircraft had been converted to RA-5C reconnaissance models, and by the end of 1965, the Vigi had become an integral part of carrier air wings operating in the Vietnam theatre. So successful was the Vigilante in the Vietnam conflict, and the missions so dangerous, that in 1968 the Navy ordered 46 additional Vigilantes three years after the original production line had been shut down. North American Aviation subsequently delivered 36 of these aircraft.

Vigilante of RVAH-7 over North Vietnam as photographed by F-4 escort.

The advanced and innovative features that made the Vigilante such an effective reconnaissance platform also proved to be its Achilles' heel. Such cutting-edge systems were in their infancy and required constant maintenance. As with the technologically advanced A-6 Intruder, Vigilante squadrons were accompanied aboard the carrier by North American Aviation and other company representatives. A "Hangar Queen" could often be found in both Vigilante and Intruder squadrons, which was an aircraft used for spare parts until regulations required it to be flown. At that time maintenance crews would scramble to make the cannibalized aircraft airworthy again. Though many technical issues were eventually remedied on both aircraft, the Vigilante remained a decidedly high-maintenance airplane that occupied a large footprint on space-conscious carriers. However, when operating efficiently there was no other Navy aircraft of the time that could provide such an effective reconnaissance platform. The expertise required to master the RA-5C, its highly hazardous role in the Vietnam theatre, and the legendary success it achieved came with a steep price. The Vigilante has the sobering distinction of having the highest loss ratio of any Navy fixed-wing aircraft during the conflict. A total of 18 RA-5C airframes were lost in the theatre, 11 of which were brought down by AAA fire.

As advanced and effective as the Vigilante was, to fully understand the capabilities and accomplishments of the aircraft, one must look at the men who flew them. It took a unique breed, those select few who could master the complex flight, navigation, reconnaissance, and electronic warfare systems.

It was the summer of 1966 when Navy Heavy Reconnaissance Squadron Seven (RVAH-7) *Peacemakers* call sign "Flare" began preparing for their second WESTPAC aboard *Enterprise*. Unlike the previous 1965-1966 cruise which began on Dixie Station off the coast of South Vietnam, the entirety of the 1966-1967 cruise would be spent on Yankee Station, flying missions over targets in North Vietnam.

The skipper of RVAH-7 at the time of deployment was Commander Robert "Bob" Lovelace, a veteran naval aviator who had earned his wings in 1948. He was known for surviving a violent ejection from a Vigilante (BuNo. 148930) while on a training flight near NAS Sanford. On 9 September 1963, he was flying with an enlisted petty officer B/N in the back seat. A major system malfunction put the aircraft out of control. Forced to eject at supersonic speed, most of his clothes were ripped from his body. Landing on a rural Florida farm, the partially clothed naval aviator sporting a face filled with blue and purple bruises knocked on the door of the house. Hastily explaining the incident and his rather odd appearance, he used their phone to call the duty officer at NAS Sanford. Both Lovelace and his B/N escaped without serious injury.

The XO of RVAH-7 was 39-year-old Commander Philip Ryan. Hailing from Goodhue, Minnesota (Pop. 500), he had been a standout multi-sport athlete in high school. After playing football for the University of Minnesota for one year he was offered a tryout with the Green Bay Packers. However, he had already accepted an appointment to USNA. At the academy he excelled in football and lacrosse, becoming captain of both teams his senior year. This resulted in him winning the Thompson Trophy, which was presented to the midshipman declared by the academy's athletic committee to have done the most during the year for the promotion of athletics. He graduated from the academy in 1950.

Commander Ryan may not have been born into a Navy family, however, he certainly did marry into one. His wife Betsey's grandfather was a Norwegian seaman who emigrated to the United States and joined the Navy. He had five sons who graduated from USNA, one son who died in a fire at an academy prep school, and one daughter who married an academy graduate. Phil's Father-in-law had been an admiral who was Naval Attaché to the Soviet Union during WWII and attended the Yalta Conference post-war.

Phil Ryan was a personable man as well as a natural-born leader. Devoutly Catholic, even-tempered, fair, thoughtful, considerate, and fiercely protective of those under his command, he was the embodiment of an officer and a gentleman. The men of RVAH-7, officers and enlisted alike, had tremendous respect and admiration for him, who in short time would become their commanding officer.

RVAH-7 RAN Lieutenant (junior grade) Jim Owen flew in the back seat with Phil, whose skilled aviating saved his life on more than one occasion. He recalls that in all the time they served together he never once heard Phil Ryan raise his voice or become visibly angry. Jim remarks:

Commander Robert Lovelace

Commander Phil Ryan

He had a look he would give you, not of anger but of disappointment. That was all it took to know he was not happy with your performance and to do better. He was a role model that the entire squadron looked up to.

A particular memory Jim savors is that of Phil's "little black book" packed with names, addresses, and phone numbers. "No matter where we flew in the world" Jim remembers, "Skipper Ryan would pull out that little book, head to a phone, and plan a visit. He had friends everywhere we went".

RAN Pete Carrothers was a young Lieutenant (junior grade) when he joined RVAH-7 in the summer of 1966. A 1963 graduate of USNA, he had served his first tour out of the academy as a navigator with VW-13 flying in a modified Constellation WV-2 Warning Star ("Willie Victor"). Patrolling the DEW line between Newfoundland, Iceland, and the Azores Islands, their job was to watch over the North Atlantic for any Soviet military activity. While Pete enjoyed the task of keeping eyes on the Russians the isolation of Newfoundland, the notoriously bad weather of the North Atlantic, and long patrol hours were not particularly to his liking. Hearing about the Vigilante program he jumped at the opportunity, even if it meant flying combat in Vietnam.

When Pete arrived at RVAH-7 he was paired with pilot Lieutenant John Sutor, who was an anomaly in the world of Vigilante squadrons. Due to the challenging complexity of the aircraft, the inherent danger of the missions, and the difficulty of landing a Vigi on a carrier, RA-5C pilots were usually senior lieutenant commanders or commanders. John was an exception to this as he was an exceptional pilot. Plucked from an A-4 Skyhawk squadron in 1964, John had already flown one Vigi tour of Vietnam with RVAH-7 when he experienced his first of two ejections.

RVAH-7 *Peacemakers* aboard *Enterprise* fall 1966
Back Row: Artlip, Clark, Sutor, Pritscher, Ryan, Davison, Fowlkes, Hamrick Front Row: Carrothers, Feldhaus, Schaefer, Joyner, Bitzberger, Owen, Boyter, Osborn, Capewell

On 15 December 1965, while flying from *Enterprise* on Dixie Station, he and his RAN were forced to eject[20] from their crippled aircraft (BuNo. 151633) and were rescued before they could be captured. Pete remembers his good fortune of being paired with John:

> I was extremely lucky to fly with John, who was the only first-tour aviator that got to fly the Vigilante. I think there were a couple more A-4 pilots selected to fly the Vigi but John was the only one I was aware of who made it through the grueling training. I could not have asked for a better pilot to fly with. John saved my life on several occasions. He is a hero to me.

RVAH-7 AIO Lieutenant Scott Reuther recalls that John, who was also the squadron safety officer, was a very dedicated pilot. When not on duty he could often be found reviewing the NATOPS manual or studying intelligence. Scott remembers "John shared a stateroom with the intelligence officers instead of with other pilots. I think we rubbed off on him as he had a deep understanding and appreciation of the work we did in IOIC".

Though heavy reconnaissance squadrons were primarily composed of naval aviators of the highest caliber, their personalities were nevertheless quite unconventional. Pete memorably describes RVAH-7 as "Alpha male extrovert personalities who were really good aviators, but to call us colorful doesn't come close to covering it".

One such personality was RAN Lieutenant John Capewell, who was well known in the squadron for spinning tall tales such as when RVAH-7 were at their homeport at NAS Sanford readying for their second deployment to North Vietnam. A mandatory AOM was called for a Saturday morning safety briefing. Each officer had to select a subject to present, applicable to their upcoming deployment to SEA. John was presenting on jungle survival and what to do if you got a snake bite. He described how he and a companion were hiking across the Southwest desert when his companion was bitten by a rattlesnake. John claimed he got out his knife, slit the wound open, and sucked the venom out. Pete had the perfect reply:

> Being a smug junior officer at the time, I just couldn't resist. I raised my hand and loudly volunteered to go up the road to *Ross Allen's Reptile Farm*, borrow a rattlesnake, come back to the meeting, and let it bite my ass so John could demonstrate this lifesaving technique to the squadron.

Following the predictable uproar of laughter, hoots, and howls Skipper Lovelace yelled "Come on guys, this is serious!" However, Pete remembers that at the same time he issued this stern admonishment he was unable to conceal the amused look on his face.

It was not unusual for flight officers to carry a higher-caliber sidearm instead of, or in addition to, the standard-issue .38 revolver. It was also not unusual for them to carry additional weapons, concealed or otherwise. John replaced his standard-issue .38 caliber revolver with a personal .357 magnum. He also flew combat missions with a three-foot-long machete strapped to his back. Pleas from others in the squadron of the danger of having a large machete coming loose during an ejection or forced landing had no impact on John. He flatly refused to fly combat without it.

Another boisterous personality was pilot Lieutenant Commander John Henry Fowlkes. The Vigilante had a fuel dump nozzle directly beneath the vertical stabilizer, between the two engine exhausts. When fuel was dumped while in afterburner hot exhaust gases from the engines would ignite the atomized fuel and produce a long flame extending behind the airplane. John was known in the squadron for his rather injudicious use of this phenomenon. Pete recalls the first time he did this on *Enterprise*:

> After catapulting from the carrier John opened the fuel dump, producing a long flame behind the aircraft. Unfortunately, he hadn't alerted the rest of us beforehand that he was going to do this, so we hadn't been ready to capture it on our cameras.
>
> John was dressed down by Enterprise Skipper Holloway and sternly warned not to do it again. Toward the end of the deployment we successfully goaded him to repeat the stunt on a night launch, this time with us topside and our cameras ready. Once again John gave a memorable show. Needless to say, Captain Holloway was incensed and quickly handed out punishment.

Pete remembers John riding the ship back to homeport in Alameda while the rest of the squadron flew home to Sanford. He was stripped of his wings and sent first to an inconsequential bombing range in south Texas, followed by orders to a remote atoll in the Aleutian Islands chain.

However, to John, whose attitude towards command authority was "What are they going to do, court-martial me and not let me fly over Hanoi?" the punishment was well worth it. At squadron reunions over the decades he has received standing ovations for his glorious act of insubordination.

Vigis dumping fuel while in afterburner.

Pilot Lieutenant Commander Frank "Hammer" Hamrick, who was the administrative officer of RVAH-7, was known for being no-nonsense. He was once interviewed on *Enterprise* during the 1966-1967 WESTPAC cruise by a reporter from a major American news magazine. When asked what he thought of the conflict he replied "It's not a very good one, but it's the only one we have".

Frank once wrote of a particular RAN he flew with. As the RAN had no visibility outside of his two small windows, which were typically closed and covered with blinds during combat missions, he was constantly querying Frank on what was going on outside the jostling aircraft. In the middle of a challenging BDA run, he again asked what was happening outside. Frank, exasperated with his often untimely inquiries, told him to open one of his blinds and take a look himself. When the RAN did he was confronted with endless streams of tracers and AAA enveloping and bursting around their aircraft. The RAN quickly closed the blind and never asked again.[21] In November 1970 Frank Hamrick became CO of RVAH-13. He is considered one of the greats to ever fly a Vigilante.

Lieutenant Commander George Clark was a highly experienced pilot. After graduating from USNA in 1956 and USNTPS in 1963, he was the primary test pilot on the Vigilante program and also performed the initial carrier qualifications for the aircraft at NAS Patuxent (Pax) River. Before joining the Vigilante program he had flown the A-4 Skyhawk from 1960-1962, completing three Mediterranean cruises aboard *Independence*. He joined RVAH-7 in 1966 where he remained for the next three years.

Initially, George was puzzled as to how he was selected to test the new supersonic RA-5C as there were other highly qualified test pilots in the program with more seniority and experience. About a week into the program a civilian engineer revealed the reason why. It was because none of the senior test pilots, fearful of the airplane, wanted to fly it.

One day during Vigilante qualifications at Pax River he was performing touch-and-go landings. After taking off and turning crosswind to downwind the aircraft's drogue parachute unexpectedly deployed, plunging his airspeed dangerously low. He tried to jettison the chute, however, that was unsuccessful. About 100 feet above the ground, with his airspeed and altitude deteriorating, George cycled the afterburner several times. This eventually burned through the shroud lines of the drogue chute, freeing the Vigilante from the excessive drag.

After surviving qualifying the Vigilante for carrier operations, as well as a combat tour of North Vietnam with RVAH-7, George almost lost his life in May 1969 at the controls of a WWII-era R4D (C-47) Skytrain in Charleston, South Carolina:

> Since most of my time was flying jets, I wanted to get some prop time in the C-47. I had a total of 10 hours in the left seat when this occurred. There was a flight going from Albany, GA to Charleston, SC. I asked my instructor if I could fly the left seat for that flight and he said yes. When we got to Charleston there was a pretty good crosswind. On landing, with the tail acting like a sail, the plane started to fishtail down the runway. I attempted to correct with rudder and throttle and the plane drifted to the right. Partway down the runway the plane made a hard swerve to the left. With that, the instructor took over. As we approached the side of the runway he frantically tried to get airborne. We got about ninety feet in the air and the plane stalled; the left wing struck the ground and the cockpit started to nose-over. The last thing I remember was putting my hands in front of my face and then I went unconscious. I was out for about 40 minutes in all. When I finally recovered some weeks later I asked for a flight physical. I passed and went back to flying for another two years!

George and Pete Carrothers were in the RAG together before being assigned to RVAH-7. They had been paired as a team until George pulled a prank that left Pete fuming:

One night George and I were scheduled for a training mission from NAS Sanford, where the RAG was stationed at the time. We were strapped in with the canopies closed, going through our respective checklists when there was a loud noise, what sounded like an explosion, from the rear of the aircraft. I also thought I saw an orange glow through the small windows I had in the back seat. On the ICS I yelled to George "What was that?"

He responded, "I think we're on fire!"

My adrenaline was already at high pucker factor[22] and I began to sense that my seat was getting hot. I yelled something about not being able to eject, and George started laughing. He said he was just pulling my leg, and that the loud noise and orange glow were from the auxiliary power unit that had backfired. I said I was done for the night and canceled the training flight.

The next day I reported the incident to the RAG commander and asked to be reassigned to another pilot. That is how I came to fly with John Sutor, whose RAN had finished his tour and departed the squadron. Bill Schaefer was then assigned to fly with George. I didn't stay angry with George for very long at all. Our families eventually became close friends.

Pilot Lieutenant Commander Greg Davison was an effervescent personality and was considered the consummate connoisseur of mischief among the officers of RVAH-7. His nickname was "Snoopy" and he had a leather helmet and a silk scarf that he wore until he got strapped into the Vigilante. Greg's motto was "No smoking 24 hours before the flight and no drinking within 50 feet of the aircraft". One of his prankish traits was flying low and supersonic while on training flights around Florida, leaving numerous windows shattered in his wake. RAN Lieutenant (junior grade) Bill Schaefer recalls a nighttime training flight with Greg:

> My first Mach-plus flight was down in the Everglades with Greg. We were hauling supersonic at treetop level when he told me to take off my glove and feel the airplane skin. It was warm and getting hot. Nice flight, but when we came back the maintenance guys were pissed. Greg forgot to retract the anti-collision light and left it in some swamp somewhere.

Though Greg is fondly remembered for his many hijinks, he is also remembered as the daring pilot who reconnoitered every inch of Route 1 in North Vietnam, from the DMZ to Hanoi while flying just a half-mile off the coastline. These missions were highly dangerous and risky however the intelligence obtained proved invaluable to the Navy and other branches of the military.

RAN Lieutenant Tom Joyner was beginning his second WESTPAC with RVAH-7. On his first deployment in 1965-1966, he flew with pilot Commander Bob "Sweet Lips" Donaldson, who acquired the nickname from his often acerbic, direct, and blunt manner of speaking. Tom remembers one night when Bob's cockpit canopy flew off when catapulting from *Enterprise:*

I heard and felt it right away and knew something wasn't right. I couldn't see forward to the cockpit but when I checked my instruments everything looked fine. After a minute or so Bob recovered himself enough to get on the ICS. In typical Sweet Lips fashion, he exclaimed "Joyner, if you punch out on me I'm gonna have your ass!"

Despite the missing canopy and gale-force winds blowing into the cockpit, Bob was able to fly the airplane back to the carrier and land safely. Tom also recalls the RAN compartment getting very warm when flying in the Vigilante, as an aircraft flying at supersonic speed generates a tremendous amount of heat:

We always wore our gloves when flying, because the skin of the aircraft and everything in the RAN compartment would get so hot it could burn your fingers if touched without a glove on. Supersonic flight and the weather in SEA were hard on the aircraft, with it common to have patches of blistering and peeling paint on the fuselage. There were cans of paint on the flight deck, and if a Vigi needed to be turned around quickly maintenance personnel would slap on a few coats to keep the skin protected. That was a bandaid until it could be properly addressed during maintenance in the hangar deck or Cubi Point.

It was with Tom in the back seat as RAN in what pilot John Sutor describes as the most terrifying experience he had while flying the Vigilante. It came not in combat, but in early September 1966 while on a terrain avoidance testing and evaluation flight in the remote and lonesome hilly hollers of northern Arkansas. Tom remembers:

> The special routes we were flying were developed by a Joint Service Terrain Avoidance Radar Effectiveness Evaluation Group. Our Vigilante (BuNo. 148933) was one of multiple U.S. military terrain avoidance, radar-equipped aircraft under effectiveness evaluation. The evaluation routes were located in the very rural hills of Arkansas, and most nights were pitch black.

Departing in darkness from England Air Force Base which was located near Alexandria, Louisiana they began their run over the prescribed route. Tom remembers that at that point in time neither man had a high degree of trust in the RA-5C terrain avoidance system capability or accuracy:

> We used our own crew safety backup procedure, which involved me monitoring the radar altimeter and calling out the reading over the ICS to John. During that flight the radar altimeter reading had been changing at moderately safe rates, giving John sufficient time to make corrective changes in altitude. Suddenly the altitude reading started to unwind at an alarming rate in a matter of mere seconds, indicating a very sudden steep incline in terrain. That is when I said to John words to the effect of "You need to pull up in afterburner now!"

And did he ever! He immediately had the plane performing like a ballistic missile. Later we joked that some poor farmer may have lost the roof off his chicken coop and probably wondered what caused all that thunder and fire in the middle of the night.

RAN Ensign John Bitzberger was the most junior officer on that cruise. He normally flew with Lieutenant Commander Lee Pritscher, who had come from the A-3 Skywarrior. John recalls "Lee was a good pilot. When we were in the air, we were like one person. We trusted each other and I know he trusted me enough to listen when I raised questions or serious concerns".

Being the most junior officer in the squadron, John was responsible for supervising a group of airman apprentices known as "coop cleaners". They were straight out of boot camp and performed basic tasks such as laundry and mess detail. John remembers a time when one of them was formally charged with theft, accused of stealing from another sailor's bunk:

> The kid who was accused was shy, had no friends on the boat, and was always getting picked on. Phil asked me to look into the matter. It turns out the missing item was something inconsequential with a value of about $1.75. I reported these findings back to Phil, who then went to Captain Holloway and persuaded him to drop the charges. Skipper Ryan taught me many lessons on leadership, such as never asking subordinates to do something you won't or haven't done yourself. He was a great leader and a man of tremendous honor.

Following the 1966-1967 WESTPAC Pritscher took John aside and urged him to apply for flight school. He did and wound up flying the A-7 Corsair II. Though he enjoyed serving under subsequent COs, he remarks that none could match Phil Ryan.

RVAH-7 AIO Scott Reuther (center, smiling) and VF-96 AIO Jim Hollarn (with glasses) in IOIC aboard *Enterprise*.

AIO Scott Reuther entered the Navy following graduation from college. After completing AOCS and flight training he flew the HUP-2/3 Retriever with HU-2, flying rescue and utility missions in the Mediterranean. He holds the distinction of the most landings (550) of any pilot on USS *Intrepid* (CV-11). He was one of the first six squadron pilots to transition to the Navy's new turbine-powered, all-weather helicopter, the UH-2A Seasprite. After a brief stint in the civilian world, he found himself back in the Navy. He attended Air Force Air Intelligence School in Colorado, where he trained as an AIO. Working in IOIC, he supervised the squadron's intelligence department which consisted of both officers and enlisted personnel. Among his staff were

Photo Interpreter Lieutenant (junior grade) Jimmy Hassell, Photo Interpreter Lieutenant (junior grade) Barry Humphrey, Electronic Evaluation Specialist Lieutenant (junior grade) Lew Thomas, and Chief Targeting Officer Warrant Officer Bobby Lanier. Scott proudly boasts that his analysts were so astute they could tell which pilot and RAN had taken which photographs, and what kind of day they were having. He also recalls that John and Pete were often assigned the most dangerous and critical missions as they were considered by the intelligence staff in IOIC to be the top team who would always get the job done. On hearing this Pete stated:

> When I heard what Scott said about assigning the "must-have" missions to us I called John. I told him if I had known that at the time I would have opted to fly with the guy at the bottom of the list who only got the easy missions!

Scott echoes Pete's sentiments regarding the composition of the squadron, likening them to the outcasts of the fabled WWII-era USMC attack squadron VMA-214, the *Black Sheep:*

> As far as naval aviators, we had some of the finest in the fleet. A team of men who could get the job done under the most challenging and adverse of circumstances. However, this would hardly be evident when looking at the individual personalities of the squadron. RVAH-7 ran the gamut from the straight-laced John Sutor and Frank Hamrick to the irreverent Greg Davison and defiant John Fowlkes, and everything in between.

Under Skipper Ryan's leadership, this veritable mash-up of personalities flourished, becoming an outstanding squadron and close-knit family in the process. This is because he treated us all with value, dignity, and respect. Nobody could ask for a better commanding officer. He inspired us, motivated us, and was the role model we aspired to be. A fine officer and an outstanding man.

In late September 1966, in preparation for their November departure to Yankee Station, the squadron was ordered to the west coast for ORE with *Enterprise* off the coast of California. On Saturday 24 September the flight of six Vigilantes, led by XO Phil Ryan, departed from NAS Sanford to NAS Alameda in the bay area of California. It was late morning by the time they arrived at Carswell AFB near Ft. Worth, Texas to refuel their tanks for the next leg of their journey to Nellis AFB near Las Vegas, Nevada. Across town at Cotton Bowl stadium in Dallas 58,000 energized fans packed the stands, eager for the kickoff of the football game between home team Southern Methodist University (SMU) and visiting Navy. While fuel trucks were filling the thirsty Vigilantes Phil Ryan and his RAN Jim Owen were performing pre-flight duties when Pete approached.

"Say XO, did you know SMU is playing Navy at the Cotton Bowl today?" Pete asked. Thinking for a moment before flashing a wry smile, Phil replied "Are you thinking what I am?"

"Wouldn't it be a hoot to do a formation flyover before kickoff?" Pete rhetorically asked his fellow Naval Academy alumni. After a glance at his watch Phil, along with Jim, began contacting various ATC agencies. After making their requests they resumed preparations for their departure to Nellis.

It was nearly an hour later when the squadron began taxiing out from the ramp. The XO had given it his best shot, but time was running out and the squadron had a schedule to keep. As the aircraft were holding short of the runway waiting for takeoff the radio crackled to life with a welcome message. It was Dallas ATC with their clearance, coupled with a simple request to keep their speed subsonic along with a warning about a radio antenna in the vicinity of the stadium.

Once airborne ATC vectored the flight to the Cotton Bowl, only for them to miss the mark laterally by about half a mile. Kickoff was delayed while the aircraft lined up for another pass, this time on the money and right over the stadium. The crowd erupted in thunderous cheers as the sleek, supersonic Navy reconnaissance jets roared overhead in tight formation. Cleverly, the squadron managed to score points for the Navy before the game had even begun.

Later, approaching Las Vegas, the aircraft were breathing fumes due to their Cotton Bowl escapade, which had consumed a considerable amount of fuel.[23] However, temptation once again proved too great for Phil and his men. Lining up in tight formation they made a thundering low pass over the famed Las Vegas Strip, giving both residents and tourists alike a thrill they were unlikely to ever forget.

During ORE with *Enterprise* John and Pete had the first of many close calls that almost cost them their lives. On 6 October, they catapulted in their Vigi (BuNo. 149288) off *Enterprise* for nighttime reconnaissance training off the coast of Southern California. They were being sequenced for return to the carrier along with a Douglas KA-3B Skywarrior tanker (BuNo. 147650) call sign "Folder 3" when disaster struck.

As the flight progressed the weather deteriorated, resulting in a low overcast layer with limited visibility and choppy waters. Tragically, the pilot of the KA-3B was on his first tour, with little experience flying in such challenging weather conditions. Why he was permitted to fly in that environment, one that would challenge even the most seasoned of pilots, is unknown.

*Enterprise* advised John and Pete that they were experiencing difficulty contacting the tanker and requested that they make radio contact directly. Though the Vigi was also unable to make radio contact, John and Pete did establish visual contact with the crew, exchanging hand signals. The Vigi initially took the lead, however, it was apparent that the KA-3B was having difficulty staying in formation. At this point John and Pete let the Skywarrior take the lead, tucking up under their wing, and providing some relief for the crew. It was close to midnight when the aircraft were vectored for return to the carrier.

Entering an overcast cloud layer at 1,700 feet above the water, some ten miles behind *Enterprise,* John and Pete were configuring the aircraft for landing when suddenly and without warning the Vigilante and Skywarrior violently collided mid-air. The Vigi rolled sharply to the left and the nose dropped precipitously below the horizon. John and Pete ejected, landing unharmed in the dark Pacific Ocean, a solid cloud layer hovering over them.

Landing in the water some 300 yards from John, Pete inadvertently cut the lanyard attached to his life raft, which drifted away in the ocean current. John began swimming and eventually found Pete, exclaiming "Where the hell have you been? Nancy is going to kill me!" Before leaving for ORE Pete's wife Nancy made John promise to take care of him, explaining "He's all I have". That was a promise John Sutor had no intention of breaking.

They took turns in and out of the remaining life raft for several hours until they heard rescue helicopters, at which time Pete began firing flares. They were picked up by a helicopter from *Enterprise* and returned to the carrier. A SAR operation by the Navy and Coast Guard for the crewmen of Folder 3 found no survivors. Pilot Lieutenant Deighton Hunt, B/N Ensign Carroll Gibson, and AO1 Melvin Colby perished.

After returning to NAS Sanford for one month RVAH-7 once again departed for the west coast and *Enterprise*, joining nine other squadrons that composed CVW-9, part of CTF-77. The carrier departed NAS Alameda on 19 November and, after short stays in Pearl Harbor and Subic Bay in the Philippines, began flying low-level reconnaissance missions over targets in North Vietnam on 18 December.

A week after commencing combat operations RVAH-7 and *Enterprise* received a small respite for Christmas, the second consecutive year the carrier had been on WESTPAC during the holiday. Cardinal Francis Spellman, the Archbishop of New York, came aboard to greet the men and conduct mass. He was accompanied by actresses Tippy Hedren and Diane McBane, who visited the ready room of RVAH-7 and met the squadron. Tom Joyner memorialized the holiday season on *Enterprise* with a home movie camera, which shows the crew doing their best to celebrate the holiday far away from home on Yankee Station.

After one month on the line, *Enterprise* steamed back to Subic Bay, taking a break from combat operations. CO Bob Lovelace left the squadron at that time and was replaced by Phil Ryan as CO with Commander Bill "Windy" Winberg assuming the role of XO. RAN Tom Joyner also departed the squadron, having served with the *Peacemakers* for three years.

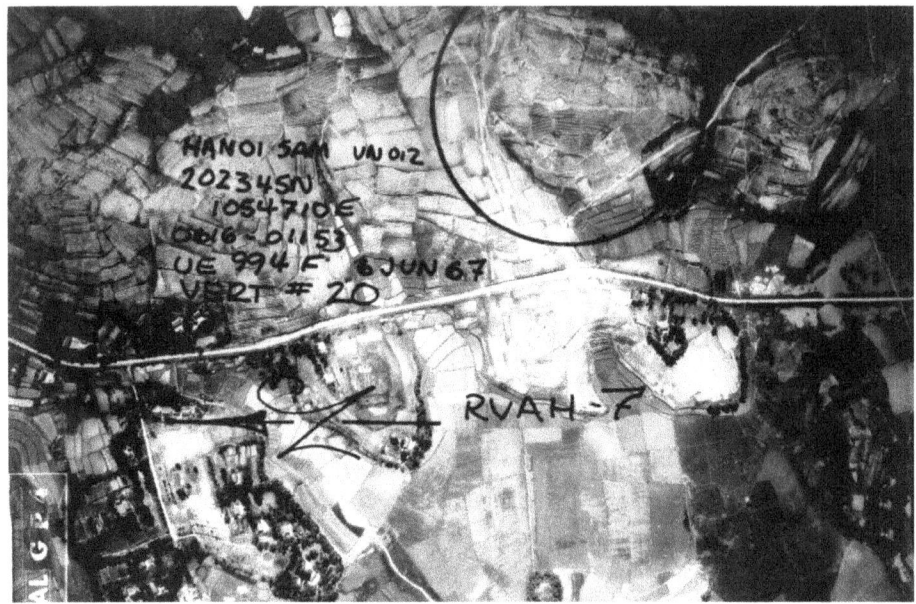

Hanoi SA-2 SAM site (circled in black).

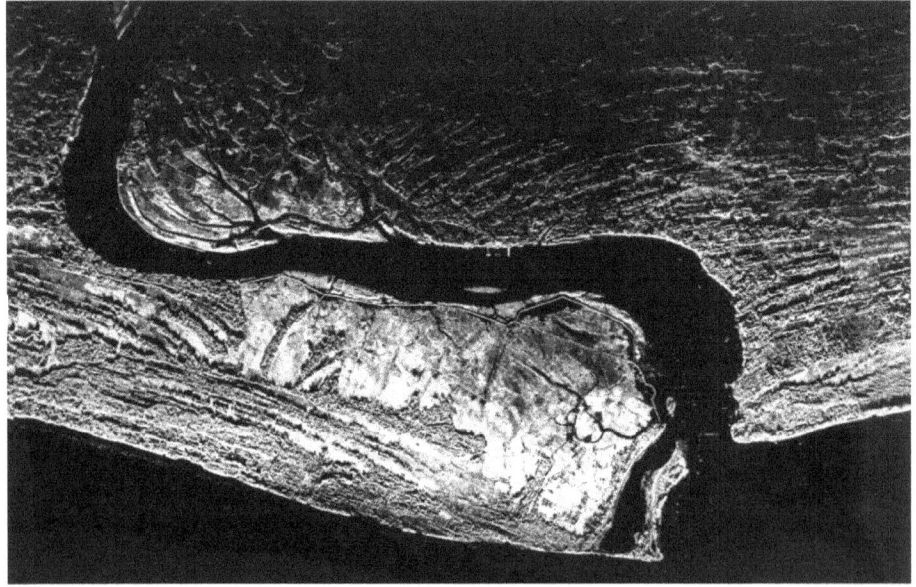

SLAR photo of Cam River from Tonkin Gulf to Haiphong.

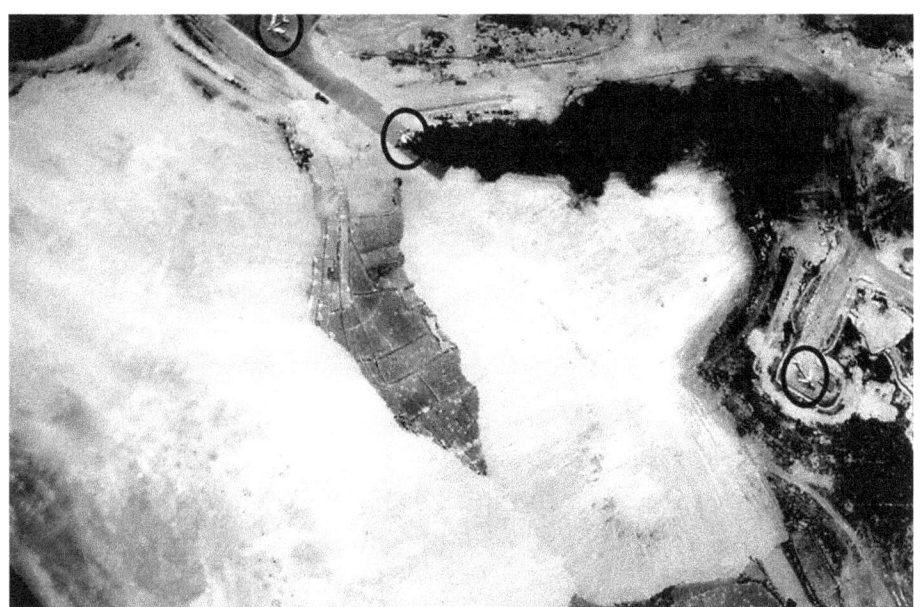

VPAF MiG-17s at VPAF airfield. The burning "smudge pot" at top was an attempt to protect the aircraft from attack.

Recon photo of VPAF airfield. MiG-17s are clearly visible, one with a smudge pot.

Carriers on Yankee Station would typically spend a month on the line before steaming to a port of call for crew liberty, maintenance, and other purposes. The sprawling Navy facilities in and around Subic Bay in the Philippines were the most common port visited between line periods. While some officers engaged in benign activities others ventured to the bizarre and wild area of Olongapo, where Shore Patrol maintained a constant vigil. Another popular recreational hotspot in the Subic Bay area was the Cubi Point O-Club. The club had a long, storied history, and on the walls were dozens of plaques made of wood from local Monkey Pod trees, each one representing a squadron that frequented the club. Some of these plaques dated back several decades to WWII.

Though chronicled within the context of RVAH-7, the capers, and romps, that made the club notable amongst generations of naval aviators can easily be ascribed to any officer that walked through its fabled doors. Junior officers generally preferred this club as it was only open when a carrier was in port and was the only O-Club that allowed flight suit attire. It was also the only club senior and flag officers tended to avoid due to the well-earned rambunctious and wild reputation. A beer cost a dime, and pitchers ran about a dollar. There was but one rule at Cubi O-Club, and that was there were no rules. Pete remembers:

> Fights were fairly common, particularly when the club was "invaded" by Marines. Despite this, management was pretty tolerant of our craziness. If they weren't I don't think they would have had many customers. The only time I can remember anyone getting kicked out was one night when a special forces guy climbed into the attic and was stabbing a machete down through the ceiling.

Although the passage of many decades has obscured numerous facets of the notorious club, some details have survived. To get to the club one would have to take a jitney up a winding road, high up on a hill through thick jungles with monkeys swinging from the trees. The main club was a building consisting of three levels, separated by stairs. A dance floor occupied the first level, a lounge on the second level, and a bar on the third. There were numerous large, glass windows throughout the club, several of which overlooked the massive bay with a view of the carriers tied up there. These windows were often broken by objects or people being thrown through them, so often that the price of replacement was well-known amongst repeat customers. Some recall the posting of a price list in the club detailing the costs for broken windows, tables, chairs, glasses, and so forth.

There is a story of a window being broken by a rowdy squadron, with a bill sent to the air wing they were attached to. A check was received for double the amount, to cover not only the current replacement but a future replacement as well that was sure to be needed. Another incident recalls an airman who was wrongly blamed for a broken window. His pleas of innocence fell on deaf ears, resulting in him being forced by his command to pay up. The next time he was in the club he picked up a chair and threw it through one of the windows, proclaiming to the angry proprietor "Now we're even!"

A reputed incident involving these vulnerable, endangered windows was when an admiral and his family were dining at a reserved table, the "Captain's Table" next to a window with a view of the bay. An inebriated aviator thought it a good idea to hang from the roofline in front of the window and bare his bottom to the unsuspecting party. The man lost his grip and crashed through the window, ending up on the admiral's tabletop.

Within the lounge area of the club were numerous long, rectangular tables. Once evening gave way to night a squadron would claim and occupy one or more of these tables as their own. After the point in which inebriation reached a rich saturation and the tabletops packed with bottles and glasses, flight operations would begin. Each squadron had differing criteria, however, the basis was essentially the same. An aviator would be seated in a chair, which was fitted with casters. Though the chair was often occupied by a nugget as a right of passage, many a veteran aviator also took a turn. They would be assigned an aircraft to "fly" which was usually the same as what the squadron was flying, though sometimes another aircraft such as the Skyraider would be selected for variety. The fearless aviator would be required to properly and realistically emulate the sounds of that aircraft, including afterburner, if so equipped.

To simulate a catapult shot the chair-bound subject would be launched across the flight of stairs that separated the dance floor and lounge. If still upright, conscious, and without incapacitating injury he would then be wheeled around the lounge by his squadron mates, with tabletops representing the flight deck of a carrier. He would come into the break, downwind, base followed by final approach. When the chair reached the table he would be picked up and hurled across the surface, with numerous bottles and glasses "arresting" his landing. The goal was to knock down as many bottles and glasses as possible. The landings were graded as they were on the carrier, with "OK 3" being a passing grade. If a passing grade was not earned the aviator may be allowed a second attempt, depending on the quorum and condition of his squadron mates. There were often multiple flight ops occurring at the same time, leading to many collisions which surely only added to the realism the men were striving so ardently to create.

An annex was built circa 1968, just down the hill from the main club. There was a noticeable absence of furniture and minimal presence of glass in this concrete block annex, a purposeful strategy to negate damages and injuries. Within the annex an attraction was added, modeled on the "Dilbert Dunker" used for water egress training in Pensacola. It is quite possible this addition was to replace the drunken flight ops at the main club that had left many aviators, tables, and chairs bruised or broken. The dunker consisted of a disused cockpit, some remember as being from an F-8, mounted on a horizontal track. At the end of the track were two barn-type doors, beyond which was a water tank. For a paltry fee of $2 an occupant would take their place in the cockpit and render a salute. Compressed air would then send the cockpit screaming down the track to the doors and water tank. If the occupant pulled a lever at just the right time a hook would catch a cable, and the cockpit would stop short of the tank. If pulling of the lever was too early or late the unfortunate victim would end up neck-deep in water, much to the delight of the onlookers. Those few fortunate enough to catch the wire were rewarded with a bottle of champagne and their name enshrined on a plaque.

The most notorious Cubi O-Club story of RVAH-7 is when five squadron officers (who wish to remain anonymous) were at the club consuming countless rounds of alcoholic concoctions. At some point, they drunkenly decided to crash the evening USO show at the main Subic Bay O-Club, even though they were in flight suits and khakis, while the main club required formal wear. After piling in a jitney and making their way down the hill they arrived at the club just before the start of the show. They discovered all the seats were taken, except a front-row table that had a "Reserved" sign sitting atop it. Undaunted, the men sauntered to the unoccupied table and seated themselves.

The show began with a well-known American entertainer of the time, in a red sequined dress, walking on stage to great applause. As she began singing one of the group hopped on stage, commandeered the microphone, and began his own drunken, impromptu show in front of numerous flag and senior officers.

In no time a young, clean-cut American man wearing a barong approached. He told them he was a member of the admiral's staff, the same admiral whose reserved table they had taken. The young man informed them they needed to get their stage-bound friend and get out. Leading the boisterous group to the lobby, they demanded his identity. He replied that his name was Captain Jack Frost. They were greatly amused by this, which prompted the drunken officer who had taken the stage to loudly and belligerently proclaim his rank as a lieutenant commander, which outranked a Marine Corps captain by one grade. Captain Frost cooly replied that he was not a Marine captain but in fact, a full Navy captain which outranked a lieutenant commander by two grades. Doubting the baby face was a full Navy captain named Jack Frost the group demanded he produce his identification. After proving both the Shore Patrol appeared, who escorted the men outside and helped load them into a jitney for the short journey back to *Enterprise*.

The five men returned to the carrier and discovered their hijinks had been reported to Skipper Holloway. He chewed them all out, with the stage-bound man receiving the brunt of the tongue-lashing. Though in conventional times the punishment would have surely put one or all "in hack" the fact that *Enterprise* was leaving the following day for Yankee Station, and the squadron could not afford to lose any men, the matter was quickly dropped.

Resuming combat in late January, John and Pete were flying a mission shortly after the annual Tet ceasefire began. The agreement between the two adversaries was that reconnaissance flights over North Vietnam could continue during Tet, however, no attacks were to take place during the lull. Pete remembers a particular reconnaissance mission below Thanh Hoa during this ceasefire:

> All my sensors indicated that for once nothing was tracking us. So we slowed down and became sightseers, just cruising around while flying at 250 knots. I remember flying over a lake and seeing North Vietnamese on a sampan waving to us, and we wagged our wings in return. John suggested we take advantage of the lull and fly up to a small island off the coast of Haiphong. We had picked the island as the IP for our upcoming flight there once the ceasefire had ended. The plan was for us to approach from the sea and use a small karst mountain ridge that lies east and west and stay behind that until we got abreast of Haiphong. We would then pop up and come to the city from the north, where they would never expect us to come from. That would give us the advantage of heading out to sea as we crossed over the city.
>
> Seizing on the temporary ceasefire we flew up there and got down on the deck as John liked to make his run into the IP at 50 feet above ground level. In the Tonkin Gulf, we would fly so low we would get sea spray on the viewfinder of the video camera on the forward belly of the fuselage. I can tell you, flying 50 feet over the water in full afterburner doing Mach-plus in the back seat was a real pucker factor. The beauty of flying with John was

that we almost always got total surprise when we popped up feet dry. This minimized the time they had to get ready and we didn't get fired at as much as other crews did. We would stay well below North Vietnamese search radar before popping up to 3,500 feet in maximum afterburner. John wanted to make sure I could pick up the island with my radar at a low altitude.

We found the island and it looked uninhabited, a grassy knoll and kind of swampy. A couple of days later we discovered the island was filled with well-camouflaged air defenses. That route was our preferred way of getting into Haiphong and Hanoi, but the risk was it was close to Chinese airspace and we were under strict orders not to penetrate it. Doing so risked a court-martial, or worse, being shot down by the Chinese.

Although John Sutor was one of the more earnest pilots in Heavy 7, he nevertheless enjoyed engaging other members of the squadron in pranks. Pete remembers occasions when he and John would be flying over the gulf at the same time as pilot John Fowlkes and RAN Lieutenant John Osborne. John would ask Pete to locate them, then would quickly creep up on their six o'clock blind spot. Unaware that another aircraft was right on their tail Pete would key the radio microphone and emulate the sounds of a firing machine gun. The two would "strafe" the other aircraft from starboard to port, leaving Fowlkes and Osborne needing clean flight suits upon their return to *Enterprise*.

On 12 February pilot, Commander Don Jarvis and RAN Lieutenant (junior grade) Paul Artlip were performing a nighttime SLAR mission off the coast near Long Chau, 30 miles northeast of Thanh Hoa.

Suddenly heavy AAA fire raked their Vigilante (BuNo. 151623), call sign "Flare 105," crippling it. Ejecting outside the envelope, Jarvis broke the arm restraining straps of the ejection seat which caused his upper extremities to windmill, tearing numerous cartilage. A rescue SAR was immediately launched with numerous Navy and Air Force aircraft. During the rescue operation, a North Vietnamese junk began approaching the men in the water. An orbiting Phantom of VF-96 from *Enterprise* fired an AIM-7 Sparrow missile at the vessel. The missile missed the target but the junk received the message and quickly left the scene. Don was picked up by an Air Force HU-16 Albatross amphibian, while Paul was picked up by a Navy SH-3 Sea King helicopter. Pete recalls:

> When Paul got back to the ship he was worse for wear. His flight suit, integrated harness, and pressure suit had been torn off and he had white strap marks where his parachute harness crossed his chest. His flight boots were about the only thing left intact.

Artlip suffered injuries that kept him grounded for several weeks. Jarvis sustained injuries that ended his flying career with the Navy. He never returned to active flight status but ended up in the U-2 program which shared several reconnaissance components with the Vigi.

On 4 February John and Pete were streaking over the eastern lowlands of North Vietnam, heading to the coast and feet wet back to *Enterprise* after another post-BDA mission. As the aircraft flew south of Thanh Hoa the AN/APR-27 scope in front of Pete briefly lit up, with accompanying audio warble tones in his headset, indicating a North

Vietnamese SAM radar had painted their aircraft. As quickly as the warning came it disappeared.

SA-2 SAM site near Thanh Hoa. The site was targeted and destroyed several days after this photo was taken.

The North Vietnamese operators (or perhaps Soviet trainers) of the newly operative SAM installation had made a deadly mistake. Perhaps testing their systems before becoming fully operational, the timing for them could not have been worse. John banked the aircraft to the north to investigate the source. With their cameras firing the SA-2 site was spotted, the distinct geometric shape unmistakably carved out of the surrounding vegetation. These missiles were particularly threatening as they were along the route commonly flown by Yankee Station aircraft during ingress and egress. Not long after returning to *Enterprise* the Thanh Hoa SAM site appeared on the target list and within a few days it was destroyed, eliminating a deadly threat.

The air war in North Vietnam began to markedly escalate in February and March when President Johnson began authorizing attacks on key targets within central Hanoi and Haiphong, targets that had been previously off-limits. As these new targets were in areas of civilian population the Navy utilized, for the first time, precision-guided AGM-12 Bullpup missiles and AGM-63 Walleye glide bombs. A-4 Skyhawks of VA-212 operating from *Bon Homme Richard* debuted the Walleye, utilizing it on a successful mission against the Sam Son Army Barracks on 11 March. A little over two months later VA-212 once again used the Walleye to obliterate the Bac Giang TPP northeast of Hanoi.

Pete remembers a day that spring when civilian defense contractors came aboard *Enterprise*. They immediately headed to the hangar deck where an A-4 had been concealed behind shrouds. Soon after John and Pete were in IOIC being briefed, along with an A-4 pilot, for a Bullpup attack on one of Haiphong's thermal power plants. Hanoi and Haiphong were located within Route Packages 6A and 6B, respectively. These were the most dangerous zones of operation for American aircraft operating in the north. The cities were guarded by 23mm, 37mm, and 57mm AAA guns, as well as 85mm and 100mm. There were also SA-2 SAM missile batteries circling the cities, making them at the time some of the most heavily defended in the world.

The mission was successful, with the A-4 pilot reporting that he guided the weapon right through a window. However, when John and Pete flew over the target post-BDA the building was still intact, seemingly the same as it was in their pre-assessment run. When the film was developed and analyzed in IOIC the photos confirmed that the exterior of the building looked the same, with no apparent damage. The A-4 pilot was flabbergasted and swore a direct hit on the target.

Scott Reuther and his intelligence team were also baffled. As they were considering that the munition may have been a dud PT2 Carl Ruby made a startling discovery. Using SCV he noticed that the sides of the building were bulged, which had not been apparent when viewing the photograph in two dimensions. It was subsequently confirmed by intelligence sources that while the walls of the building still stood, the components inside had been destroyed. It was theorized that as the doors and windows of the building had been wide open at the time of the attack, much of the blast wave was able to escape. This left the structure relatively intact while destroying everything on the inside.

Post-BDA photo of Haiphong Thermal Power Plant West. It is still on fire from the preceding attack.

Vigilante flown by Pritscher and Bitzberg hit by AAA March 1967.

In early March pilot Lee Pritscher and RAN John Bitzberger had an uncomfortably close call. They were on a road and river reconnaissance mission to Vinh, another heavily defended target seven miles west of the coast. John remembers:

> We were traveling at Mach 1.1 (837 MPH) at 3,500 feet on our run. We did the route and were feet wet when our F-4 escort came alongside and told us we were hit. There were only one or two spots on the Vigilante where a hit will not cause serious, if not debilitating damage. We were hit in one of those places, the last two feet of the horizontal stabilizer, an area with nothing but aluminum honeycomb between the upper and lower skins. Everywhere else on the bird had critical components such as fuel and hydraulic lines. Lee skillfully got us back to the carrier in one piece, where they cleared the deck and let us land first.

In the next letter John wrote home to his family he described the events that day. His Father wrote back, asking him to refrain from any more war stories. His Mother had nearly fainted.

On 9 March a Vigilante (BuNo. 151627) of RVAH-13 call sign "Flint River 605" from *Kitty Hawk* was lost near Long Chau while on a coastal photoreconnaissance mission. Paralleling the coastline one-quarter mile offshore, and just 350 feet above the ground, their F-4 escort of VF-213 *Black Lions* call sign "Black Lion 106" spotted small-arms fire coming from the shore. Shortly thereafter the Vigi burst into flames. The pilot and RAN ejected and came down in waist-deep water about 200 yards offshore. As their F-4 escort was calling rescue forces NVA began wading out into the water to capture the men.

Trying to protect the two until a SAR helicopter arrived, A-1 Skyraiders of VA-52 *Knight Riders* call sign "Viceroy" from USS *Ticonderoga* (CVA-14), also known as "*Tico*," made 20mm strafing runs. While the Skyraiders plied their trade a *Black Lion* escort flew several low passes and fired a Sparrow missile that impacted near the NVA troops. While this caused most of them to retreat to shore, two persevered and apprehended RAN Lieutenant (junior grade) Francis "Frank" Prendergast, quickly taking possession of his standard-issue .38 caliber revolver sidearm. However, the two failed to frisk him for additional weapons. As they marched him back to shore the guard armed with an AK-47 would duck under the water every time a pass was made by a Phantom or Skyraider. Seizing upon a moment when the guard was submerged, he pulled out a concealed .25 semi-automatic pistol he had hidden in his flight suit. The shocked second guard tried using the .38 revolver he had seized from him, however, the cylinder was empty from Prendergast firing it into the air when he splashed down, trying to get the attention of their escort. He shot the soldier between the eyes, killing him. When the AK-wielding guard emerged from under the water Prendergast attacked, beating him on the head and taking his rifle, which he threw in the water.

The fugitive aviator spotted a Navy Sea King rescue helicopter call sign "Loose Foot" inbound to his position and made a run for it along a sandbar. However the guard recovered, found his weapon, and gave chase. The guard, about a hundred yards behind, shot several times at him, with Frank returning fire. The Sea King helicopter swooped in broadside and unleashed a torrent of M-61 machine gun fire on the guard, killing him. One of the SAR aircraft believes they saw pilot Commander Charles "Charlie" Putnam alive on a sandbar near the beach, however, that sighting is inconclusive. The arrival of North Vietnamese reinforcements precluded another rescue attempt and Charlie Putnam was eventually declared KIA.[24]

Prendergast has the distinction of being the only American POW captured and escaped from North Vietnamese captivity (on the same day, no less). As an escaped POW who had killed an NVA soldier, he was not allowed to resume flying combat in theatre. If he were captured again there is a likelihood he would have been summarily executed. He instead returned stateside, went to Navy flight school, and became a jet pilot.

On another occasion, John and Pete were on a "back door" approach to the heavily defended port city of Haiphong for a reconnaissance mission. An 85mm AAA shell exploded close to the empennage of the Vigilante, causing both engines to flame out and lose thrust. Pete clearly remembers the dramatic incident:

> We had just popped up from 50 feet at our IP and had hit maximum afterburner when the shell exploded and all hell broke loose. Our Phantom escort yelled "You're hit, you're on fire! Eject!" The plane was in ground effect and wasn't flying, just wallowing with the nose up and holding barely above the water. I put my hands on the ejection ring above my helmet and on the ICS I told John it was time to punch out. He calmly replied, "Hold on, I'm working on it".
>
> Meanwhile, our F-4 escort kept telling us we were on fire and to eject. Finally, John got one of the afterburners lit and we started making some headway. Shortly after he got the other afterburner lit, picked up some speed, and got the hell out of there. We were both too shaken up to continue the mission to Haiphong. When we got back to the ship John downplayed the event as a hard afterburner light-off, where kerosene builds up and finally ignites. This produces black smoke which the F-4 thought was a fire. I'm not convinced that the 85mm shell that exploded right behind us didn't cause the flameout. John assured me that wasn't the case. All I know is 99% of the pilots in that scenario would have ejected. That is another incident where John likely saved my life.

124

On 14 March *Enterprise* took another break from the line and steamed to Hong Kong. Pete and fellow RAN Lieutenant (junior grade) Bill Feldhaus, with whom he shared a stateroom on that cruise, went ashore and bought matching green sport coats from one of the city's fine tailors. They jokingly signed a pact that loosely read "To Whom It May Concern: If either one of us buys the farm the other guy gets both sport coats".

The following year RVAH-7 was on a Mediterranean cruise aboard *Independence*. On 14 May 1968 pilot Lee Pritscher and Bill were catapulting for a night mission when FOD was ingested into the engines of their Vigilante (BuNo. 147854), resulting in catastrophic failure. The two men ejected horizontally and miraculously survived without serious injury. They were picked up by the destroyer USS *Charles P. Cecil* (DD-835). Shortly after, Pete was called to the bridge of *Independence* to help decipher a strange semaphore message coming from *Cecil*. The message was from Bill and read "Tell Lieutenant Carrothers to keep his hands off my sport coat!"

While doubtful Pritscher and Feldhaus had the time to consult it, in case of emergency Vigilante crews had a book of checklists to follow, a small, loose-ring booklet they called the "Panic Pal". It detailed what steps to take in case of an engine fire, hydraulics failure, electrical system malfunction, and so forth. One of the known quirks of the RA-5C was the CG being too far forward if fuel was unevenly distributed. If not caught by the crew before launch the aircraft would pitch down with little to no time or altitude to recover. According to the Panic Pal, if a forward CG caused the aircraft to pitch down the first action to take was "RAN Eject" which would shift the CG aft, but could also very well be the end of the road for the RAN. Most redacted that solution, instead boldly printing on the Forward CG page "FORGET IT!"

On one particular day, pilot George Clark and RAN Lieutenant Bill Schaefer were providentially able to lend a hand on a SAR mission. Egressing out of North Vietnam and going feet wet after flying a reconnaissance sortie, the two were headed back to *Enterprise* with their valuable intelligence data. As they crossed the coastline George scanned his instruments carefully, ensuring the engines and other systems were within limits and operating properly. Glancing up over the cockpit canopy, dead ahead was a Navy A-4 Skyhawk pilot who had ejected from his crippled aircraft and was descending by parachute into the Tonkin Gulf. The pilot was so close that George had to take evasive maneuvers to avoid hitting and killing him.

Exclaiming a few choice expletives, he banked the Vigilante and circled, watching the man splash down in the gulf while North Vietnamese boats began heading towards the pilot, who was struggling in the water. Being a reconnaissance aircraft the Vigilante was unarmed, so their options were limited. After calling on the radio for rescue George came up with a plan to buy the pilot some time. Making a steep turn while sharply descending to just 50 feet above the water, he lined up the nose of the Vigilante with the approaching enemy flotilla. Throttling up to afterburner, the Vigi roared over the boats at supersonic speed, producing a thundering sonic boom. Glancing back he could see that most of the boat crews, assuming the Vigi was armed and making an attack run, had jumped into the water. George followed up with a second supersonic pass over the enemy. No closer to the downed airman, their fate was sealed when members of the pilot's squadron showed up on-scene. Strafing the enemy boats with 20mm fire, the threat was eliminated and the pilot was successfully rescued by Navy helicopter.

Photo of shore installations in Haiphong.

On 30 April John and Pete were flying a daytime photo reconnaissance mission over the heavily defended areas of Nam Dinh, Phu Ly, and Ninh Vinh. This was the southern tip of the "Iron Triangle" with the two northern points being Hanoi and Haiphong. The photographs were urgently needed to provide pre-strike information for future attacks against these valuable targets, which included railway infrastructure and industrial factories. Also located in Nam Dinh was the heavily defended headquarters of the NVA.

The two men had barely begun the run over the target when a heavy barrage of 85mm AAA opened up on their aircraft. Constantly jinking to evade no less than twenty AAA bursts, they successfully obtained a complete set of photographs. Landing back on *Enterprise* they observed AAA damage on the belly of the aircraft, including a destroyed radar altimeter. Of the mission, *Enterprise* log book notes:

Carrothers skillfully operated the complex radar, navigation, and reconnaissance equipment, and the success of the mission under extremely hazardous conditions was due to careful planning and personal courage by both men, for which Carrothers was later awarded the DFC.

On 1 May pilot Lee Pritscher and RAN John Bitzberger were preparing for a reconnaissance flight. John remembers that as he was getting strapped into the aircraft Commander Ryan ran out on the flight deck and told him to climb down and sign his promotion papers to lieutenant (junior grade). He didn't want John's wife to get shortchanged on benefits should something happen to him. It was a thoughtful act and perfectly indicative of the kind of man and leader Skipper Ryan was.

John clearly remembers that mission to Kep Airfield near Hanoi. The night before he was watching the movie *The Great Race* with other members of the squadron when the phone rang:

Lee called down and I left for IOIC where we planned the mission. The next day, not long after taking off, both our radar and inertial navigation failed! I even had to run the cameras manually. Thankfully we had planned everything on paper charts with time and distances which we used as a backup. When we got to Kep it was still hot and there was a dogfight above us between F-8s from another carrier and MiGs. As we were flying over the airfield Lee asked me to use the viewfinder and tell him when we got to the runway end. I did and he banked the Vigi hard, past 90 degrees, and we got out of there.

Hoa Binh SA-2 SAM site and missile hold area.

The month of May would see terrible losses for Navy carriers operating in the Vietnam theatre. Over the course of that month, 23 airplanes were lost, with 13 airmen taken as POWs and 11 fatalities, with the heaviest losses occurring during the last half of the month. On 19 May *Enterprise, Kitty Hawk,* and *Bon Homme Richard* suffered the worst single-day loss of Navy aircraft in the conflict, six aircraft and ten men. The losses from *Kitty Hawk* included RVAH-13 pilot Lieutenant Commander James Griffin and RAN Lieutenant Jack Walters, both of whom died shortly after ejecting over Hanoi and being taken as POWs. The post-BDA flight had initially been assigned to RVAH-7, but was changed to RVAH-13 shortly before the mission began. It was an especially painful loss for RVAH-13, who during the conflict suffered the highest loss rate of any Vigilante squadron.

On 6 June pilot Frank Hamrick and RAN John Capewell reconnoitered a large SAM installation at Van Dien, the area where Navy aircraft had taken heavy losses on 19 May. The following day the site was destroyed in an Alpha strike by aircraft from *Enterprise*. Following the attack, some ten miles away, pilot Phil Ryan and RAN Jim Owen were lining up in their Vigilante for a post-BDA flight over the former missile site. Once reaching the IP and flying 100 feet off the ground, the shutters of their cameras clicking, Phil pushed the throttles to afterburner and pulled back on the stick. The sleek aircraft instantly responded, climbing to 3,500 feet and accelerating to supersonic speed. NVA soldiers began firing a AAA gun upward in an arcing pattern, putting up a curtain of hot lead slugs. As the Vigi was very difficult to hit with AAA at such low altitude and high speed, the soldiers were hoping the aircraft would fly into the stream of AAA rounds, ingesting FOD into the engines. This would likely disable the aircraft and bring it down.

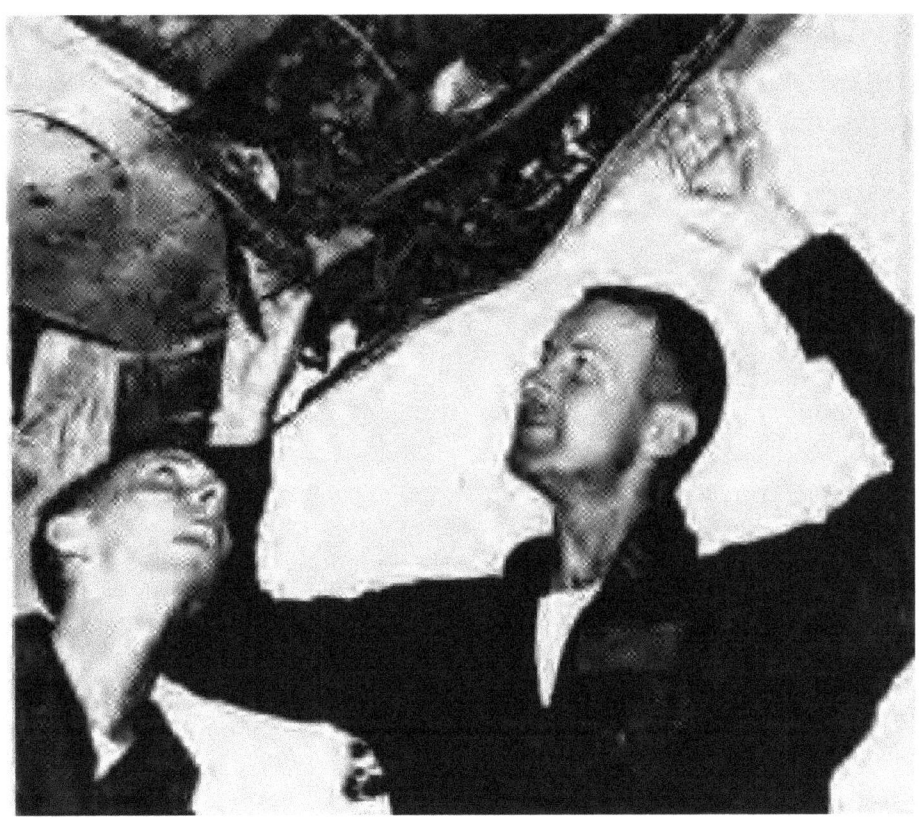

Jim Owen and Phil Ryan inspecting damage to the fuselage from AAA 6 June 1967.

Phil lined up the Vigilante for a direct pass over the columns of smoke. In the backseat, working in near total darkness, Jim ensured the cameras were functioning while simultaneously monitoring his AN/APR-27 scope for signs of enemy radar activity. As they barreled down on the site Phil spotted flashes of the AAA gun firing directly into their flight path. In a fraction of a second, he jabbed the stick over, putting the aircraft in a hard bank. Rolling out, they set a course for the coast, feet wet and *Enterprise*. Within a few minutes their F-4 Phantom escort, who had been loitering on station, appeared on their wing.

After landing on *Enterprise* Phil and Jim surveyed the exterior of the aircraft. On the belly of the fuselage, a AAA round had scored a glancing hit, destroying the radar altimeter and breaching the fuselage. Directly above sat Jim, who had come perilously close to serious injury or death. When the squadron returned to NAS Sanford after the cruise, Phil's wife Betsey was there to greet him. As Jim was describing to her the harrowing near-miss, Phil stood behind his wife, frantically motioning for Jim to change the subject, lest she know just how close to disaster they had come that day.

On 20 June *Enterprise* departed Yankee Station, concluding their 1966-1967 WESTPAC cruise. They arrived back at their homeport of NAS Alameda in San Francisco on Thursday 6 July. Scott Reuther made his way to the flight deck as the mighty nuclear carrier prepared to pass underneath the Golden Gate Bridge. He remembers:

> We arrived back in San Francisco at about midday. Skipper Holloway had us going at a pretty good clip, and there were fire boats alongside shooting red and blue water. As we approached the Golden Gate the skipper let go full blast with the ship's horn. *Enterprise* had the loudest horn afloat and it echoed throughout all of the San Francisco Bay Area. He laid on the horn for several minutes, filling me with an immense feeling of pride. Then eggs started raining down from protesters on the bridge.

RVAH-7 flew home to NAS Sanford in their Vigilantes, back to their families, having not lost a single man that cruise. Some stayed with RVAH-7 for a Mediterranean cruise on *Independence* the following year, some left for other assignments while others departed military service for civilian pursuits. John Sutor went on to fly for American Airlines, a career that lasted thirty years. For Pete Carrothers, who flew 68 combat missions that cruise, he is forever thankful to John that he was able to come home to his wife and family. In recognition of this John became godfather to one of Pete's children, a fitting tribute to the man to whom he owed his life. As for Skipper Ryan, Pete echoes the sentiments of others in the squadron. "Phil was a great skipper and a great man. I have nothing but wonderful memories of him."

Though the 1966-1967 WESTPAC was over, it was far from the end of the relationships that had been created between the men of RVAH-7 on that cruise. For them, unbreakable bonds had been formed. Bonds forged through a fight for survival in the face of withering enemy fire, on a foreign battlefield half a world away. In many ways, they had become a close-knit family. No matter how much time would pass or where they were in the world, RVAH-7 would always be a part of them.

In December 1968, nearly eighteen months after RVAH-7 returned from their second Vietnam cruise, there was one last act to be had before the curtain came down, courtesy of Greg Davison. Pete was serving on the staff of Commander Naval Air Forces Atlantic Fleet (COMNAVAIRLANT) in Norfolk, Virginia waiting for his discharge from the Navy to be processed. Greg was at NAS Albany, Georgia also awaiting his discharge from active duty. Growing impatient at the slow bureaucratic process, he decided to try and move things along.

RVAH-7 had an R-4D aircraft they used to move personnel and cargo. It was just before Christmas when Greg, having been checked out on the airplane, decided on a quick jaunt to New York City with about a dozen enlisted men from his squadron. He took the left seat while the co-pilot was an A-4 driver that he had recruited for the illicit excursion. Leaving NAS Albany they stopped at NAS Norfolk and picked up Pete.

When the group arrived at NAS Brooklyn they met John Sutor, who was by then out of the Navy and flying for American Airlines. John brought along two friends from the airline, and the party kicked off, raging into the night and early the next morning. When the time came to leave Greg and the rest were worse for wear, each nursing hangovers except John Sutor, who was a teetotaler. It also did not help that the weather that morning was IMC, with soupy fog and a low overcast ceiling. The duty officer was initially hesitant about letting the hungover motley group depart in such weather. However, Greg pulled rank and soon they were off. John stayed in New York, however, his two friends from the airline climbed aboard the R4D, asking to be dropped off in Norfolk with Pete. As the Skytrain climbed above the clouds and headed south along the eastern seaboard Pete recalls the comical scene that unfolded:

> We had climbed above the undercast when Greg came back to the cabin and said he wasn't feeling well. He asked if I would go up to the cockpit and take the left seat to keep his co-pilot buddy company. By that time we were flying VFR and on autopilot. Greg said just keep the coast in sight and wake him up when we saw Chesapeake Bay. He then went to the cabin and was soon sleeping off the late-night fun with the other snoozing revelers. After a while the A-4 co-pilot said he wasn't feeling well either, so he went back to sleep it off and one of John Sutor's airline friends came up and took the right seat. Now you have a hungover NFO and a hungover civilian, neither of

us pilots, at the controls of a military aircraft filled with hungover military personnel. If that wasn't bad enough we were also flying very close to sensitive airspace over Washington, DC.

When we saw Chesapeake Bay we woke Greg from his slumber. I mentioned to him that landing at NAS Norfolk with two "stowaway" civilians would probably not be a good idea. He agreed, landed at Norfolk Municipal Airport, taxied up to an empty gate and the three of us jumped out of the aircraft with the engines still running. He taxied back to the runway, took off, and flew back to NAS Albany, never hearing a word about it. His discharge eventually came through, and although I don't know if the escapade to New York sped up the process I doubt Greg had any regrets one way or the other.

For the next fifty years the men of RVAH-7 often gathered for reunions, happily reminiscing of their *esprit de corps* and youthful gusto, the many playful indiscretions, and streaking at the speed of sound across jungle treetops in their mighty steeds of aluminum and titanium, defiant in the face of death. It was inevitable that time and attrition would one day take a toll on their ranks. In 2018 Phil Ryan passed away in Annapolis at age 90. The death of the skipper was devastating to the surviving men and their families, who had a certain love and affection for him that can only be truly understood by those with whom he served. As to how the squadron survived a tour in North Vietnam at the height of the air war, flying 650 combat missions without a single loss of life, the squadron credits that small miracle to the leadership of Skipper Phil Ryan.

For the families of the squadron in Sanford and elsewhere, there was little doubt as to how they survived seven grueling months without their loved ones. Just as Phil had protected and looked after the men during their deployment, his wife Betsey, known as "Bets" or even "Mom" to some in the squadron, had done the same for the families back home. She was the glue that held them together. A sympathetic ear to listen, a shoulder to cry on, and a comforting presence as public support for the conflict plummeted, with the rising death toll and gruesome images splashed daily on television screens and newspapers throughout the country.

It was not just squadron reunions that brought the men back to see Phil Ryan over the years, many decades after they left the Navy. Invariably, as life's problems manifested, Phil was once again a trusted confidante and counsel, providing a sensible voice of reason with the same fatherly demeanor and gentle wisdom that had seen them through their combat deployment. Cleta Humphrey, widow of intelligence officer Lieutenant (junior grade) Barry Humphrey, echoes the heartfelt sentiments of others in the squadron where her husband proudly served:

> The Ryans were highly moral people, which was the foundation of all they did. They set the tone for family life within the squadron. They were both natural leaders and made a great team. I used to think it was by chance that the very top people were assigned to RVAH-7. Now I believe that by the Ryans valuing each person, it made everyone become the best they could be.
>
> Heavy 7 was the Navy for Barry—it was everything he thought service should be. There is a reason the group remained cohesive over the decades, and that reason is

the wonderful environment fostered by Phil and Betsey. When Barry was in deep decline in early 2021 as a result of Parkinson's Disease, a large floral arrangement arrived from the Ryan family and the remaining *Peacemakers*. They were a group with whom he had served fifty-four years previously but still felt very much a part of that family.

Scott Reuther tells a story of how during that 1966-1967 deployment Phil and Betsey often worked in tandem to keep morale up within the squadron, and at home. At some point during the cruise, Betsey took Polaroid pictures of the legs of the officer's wives. She sent the pictures to Phil, who hung them in the squadron ready room and asked the officers to pick out their wife's legs. Most of the officers selected the correct pair of legs. Phil then took pictures of all the officer's legs and sent those to Sanford for the wives to pick out their husband's legs. Nearly half of the wives picked bachelor John Sutor's legs, providing a good laugh for all.

Much the same as the men who once bravely flew Vigilantes in combat so many decades ago, the half-dozen or so surviving aircraft are scattered about the world. Several airframes are on static display in the United States, and abroad. One of these survivors stands proudly outside of what once was NAS Sanford, now Sanford-Orlando International Airport. Mounted on pylons, BuNo. 156632 is pitched slightly nose up, in a shallow bank as if gently maneuvering. Beautiful, majestic, and ever vigilant, what was designed as a nuclear bomber intended to deliver lethal munitions instead gained a new life and mythical reputation as the finest and most effective Navy reconnaissance aircraft to ever serve in combat.

---

[15] Powell, Robert. *RA-5C Vigilante Units in Combat*. Osprey Publishing 2004.

[16] Navy RA-5Cs would eventually make numerous reconnaissance fights of Cuba in *Jiffy Soda and Orange Tree* missions. Hanoi Hilton alumni Captain Gerald Coffey, USN (ret.) is one of the Vigi pilots who made these historic flights.

[17] Though the carrier-deployed McDonnell Douglas F-4 Phantom technically had a faster top speed (Mach 2.23 vs. Mach 2), that speed was only capable with a clean airframe and no external stores.

[18] Ship's company personnel numbered 3,000 men, while the air wing numbered 1,800 men.

[19] On 13 December 1960 pilot Commander Leroy Heath and B/N Lieutenant Larry Monroe established a world altitude record of 91,450 feet in an A3J Vigilante, beating the previous record by over four miles. This new record held for more than 13 years.

[20] Sutor and RAN Lieutenant (junior grade) Dresser were flying off the south coast of South Vietnam, close to the Cambodian border searching for sampans transporting arms. Flying at 3,800 feet above Vinh Cay Duong Bay, the men heard and felt a thump followed by an ominous grinding sound from the engine bays. With smoke and fumes filling the cockpit followed by a total loss of control they ejected and were rescued by an Army helicopter. The cause of the loss, whether mechanical or enemy action, remains unknown.

[21] Hamrick, Frank. *Just Call Me Frank! Memories of a Navy Jet Pilot*. CreateSpace Independent Publishing Platform 2013.

[22] "Pucker factor" was a commonly used term by combat airmen, meaning the involuntary reflexive constriction of the anus under stress or fear.

[23] In combat a Vigilante typically burned 1,000 pounds (147 gallons) of fuel per minute and had a fuel load of 25,000 pounds (3,676 gallons). A full fuel load could theoretically keep them over North Vietnam for 25 minutes, however their actual time in-country was usually 10-15 minutes.

[24] Powell, Robert. *RA-5C Vigilante Units in Combat*. Osprey Publishing 2004.

# Chapter III

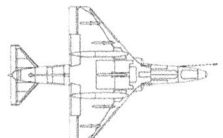

# Battle Cry of the Stingers

It was perhaps apropos that rain, fog, and mist permeated the San Francisco Bay Area that cold Saturday morning of 19 November 1966. At NAS Alameda thousands defied the inclement weather and lined the pier to bid farewell to their loved ones departing on the aircraft carrier *Enterprise* for a seven-month deployment to Vietnam as part of CTF-77. It would be the second such WESTPAC deployment for the nuclear-powered *Enterprise*, having just returned seven months earlier from its first cruise to the waters off Vietnam.

For the families of the 3,000 ship's company personnel, there were reasonable expectations that their husbands, fathers, brothers, and sons would return safely home from the protracted conflict that was beginning to ravage the social and political fabric of America. However, to the families of the men in the air wing flying combat missions, their expectations were tempered by the sobering reality of the increasingly high losses suffered by Naval Aviation since the commencement of *Operation Rolling Thunder* twenty months prior.

After a last-minute scamper of khaki and blue the brow and lines were taken up and "Big E" slowly began to drift away from the pier. The bow eventually swung north towards Treasure Island and finally west to Alcatraz, the Golden Gate, and the vast Pacific Ocean. As they passed under the perennial bridge anti-war protesters above shouted, jeered, booed, and threw eggs and red paint as the ship's horn blasted a mighty riposte. It was not long before the massive vessel became nothing more than a faint shadow on the hazy, gray horizon.

Among the combat squadrons of CVW-9 onboard *Enterprise* was Attack Squadron 113 (VA-113). The *Stingers* of VA-113 flew the Douglas A-4 Skyhawk, the indisputable ground attack workhorse of the Navy in the Vietnam conflict. The A-4 first flew in 1954 and was introduced into Navy and Marine Corps squadrons in late 1956. It was designed by famed Douglas engineer Ed Heinemann, who also designed the SBD Dauntless, A-20 Havoc, A-26 Invader, A-1 Skyraider, and A-3 Skywarrior. Among these legends of military aviation, the Skyhawk was perhaps Heinemann's most notable work.

Size and weight are critical considerations in aircraft deployed aboard an aircraft carrier. With this in mind, Heinemann delivered an airplane that was half the maximum weight specified by the Navy. The low-mounted delta wing was so compact that there was no need to fold them for carrier storage, eliminating a point of weakness in the wing structure while saving 200 pounds of weight and valuable carrier space. The undercarriage was also a stroke of ingenuity, with only the main gear wheels retracting into the wings, while the gear struts were retained in external fairings. This design allowed the main spar to span unencumbered across the entire wing. Perhaps the most innovative, and ingenious element of the airframe design were the leading-edge slats. They were designed to automatically deploy at low airspeeds simply by means of gravity and air pressure. This eliminated the need for mechanical connectors and actuators, further reducing weight, enhancing wing strength, and increasing dependability by eliminating potential points of mechanical failure.

The Skyhawk proved beyond a doubt that great things do indeed come in small packages. The aircraft was well-received by the Navy and Marine Corps, with the diminutive size drawing nicknames such as "Scooter," "Tinker Toy Bomber," and "Bantam Bomber". In an ode to its famous designer, the Skyhawk was also known as "Heinemann's Hot Rod". The small size of the A-4 along with its nimble agility, which made it a wily and difficult target for AAA, proved ideal for *Operation Rolling Thunder* missions. As such, the A-4 was lost in higher numbers than any other Navy aircraft of the conflict.[25]

VA-113 had two previous tours to Vietnam before the 1966-1967 cruise on *Enterprise*. The squadron was onboard *Kitty Hawk* as part of CVW-11 for their first deployment to Vietnamese waters in 1963-1964. It was on this cruise that the Navy lost their first aircraft of the conflict to hostile fire.[26] On 21 April 1964, a coup attempt by communist Pathet Lao forces to overthrow the pro-American Royal Laotian government prompted the United States to amass two additional carriers, USS *Constellation* (CVA-64) and *Ticonderoga* on Yankee Station,[27] roughly 120 miles off the coast of Da Nang, South Vietnam. Six weeks later, on 6 June, while flying a photoreconnaissance mission over Laos in an operation codenamed *Blue Tree*, an RF-8A Crusader (BuNo. 146823) of VFP-63 *Eyes of the Fleet* from *Kitty Hawk* piloted by Lieutenant Charles Klusmann was shot down by AAA near the village of Ban Ban. A heroic rescue effort was immediately launched by *Air America*[28] (CIA) airplanes and helicopters, however, heavy ground fire hampered their efforts. Klusmann was captured by the Pathet Lao, making him the first Navy POW of the conflict. Three months after his capture, he managed to escape to the village of Baum Long, where on 1 September he was rescued by a CIA Pilatus PC-6A from Udorn Royal Thai AFB.

Some have stated that Klusmann did not escape, rather a ransom was paid to the Pathet Lao for his release. This assertion has never been proven or disproven. Klusmann was the first of two American POWs known to successfully escape from captivity in Laos, the other being Navy Lieutenant (junior grade) Dieter Dengler.

On 1 February 1966, Dengler was flying an A1-J Skyraider (BuNo. 142031) of VA-145 *Swordsmen* from *Ranger* in an attack against a road intersection on the Ho Chi Minh Trail near the Mu Gia Pass in North Vietnam. He was shot down by AAA over Laos and quickly captured by Pathet Lao forces. After spending nearly five months in captivity Dengler, along with several other captured American airmen, miraculously escaped on 29 June 1966.

Spending the next 23 days in the jungle evading recapture, he used a white cloth to signal two USAF Skyraiders who were passing overhead, who in turn called ARRS. Air Force command elements initially turned down the A-1 pilots rescue request for an unidentified Caucasian man standing atop a rock in Laos, frantically waving a white cloth. It was only after the pilot's staunch insistence that SAR forces from Thailand were dispatched. When brought aboard the rescue helicopter the crew were shocked to discover a half-eaten snake under what remained of Dengler's tattered clothing. Of the group of escapees, only Dengler made it out alive. Over 300 American military personnel, nearly all airmen, remain missing from Laos.

The following day, 7 June, a second Navy aircraft was lost to hostile fire over Laos. An F-8D of VF-111 *Sundowners*, also from *Kitty Hawk*, was flying escort for an RF-8A *Blue Tree* photoreconnaissance flight over Laos. Commander Doyle "Bud" Lynn's aircraft (BuNo. 147064) was hit by AAA and he ejected very close to where Klusmann had the previous day. A CIA CV-2 Caribou picked up Lynn's beacon on the guard channel and the next day an agency UH-34 Choctaw was vectored in to pluck him out of his jungle hiding spot. Commander Lynn was killed less than one year later, on 27 May 1965 when his Crusader (BuNo. 148706) was hit by AAA near Vinh.

*Kitty Hawk* and VA-113 had departed Yankee Station just two weeks prior when on 5 August 1964 the Navy lost two additional aircraft, resulting in one KIA and one POW. The controversial incidents that prompted these losses are widely considered as the catalyst of what would become one of America's longest, and most protracted foreign conflicts. On the afternoon of 2 August 1964 the destroyer USS *Maddox* (DD-731) came under attack by three VPN P-4[29] torpedo attack boats in the Tonkin Gulf. Two days later, in the early morning of 4 August, a similar attack is alleged[30] to have taken place against *Maddox* as well as the destroyer USS *Turner Joy* (DD-951). Wasting no time, the administration of President Lyndon Johnson acted swiftly in responding to what became known as the Gulf of Tonkin Incident.

At 2335 EDT on 4 August (1035 5 August Indochina Time (ICT)) President Johnson addressed the nation on the incidents, informing them of the planned military response. Half a world away, west of the Paracel Islands in the South China Sea, *Operation Pierce Arrow* had already begun. Limited in scope, the operation consisted of 64 strike sorties from carriers *Ticonderoga* and *Constellation* targeting torpedo boat bases at Hon Gai, Loc Chao, Quang Khe, and Phuc Loi, as well as a POL storage depot located in Vinh.

At 1030 ICT *Ticonderoga* began launching AD-7 Skyraiders of VA-52 followed one hour and forty-five minutes later by A-4E Skyhawks of VA-56 and VA-55 *Warhorses*. Their targets were the torpedo boat bases at Quang Khe and Ben Thuy, as well as the POL storage depot in Vinh.

Fighter coverage for the aircraft from *Tico* came by way of F-8E Crusader fighters of VF-51 *Screaming Eagles* and VF-53 *Iron Angels*. At 1200 ICT *Constellation* began launching A-1H Skyraiders of VA-145. Ninety minutes later the carrier began launching A-4C Skyhawks of VA-144 *Roadrunners* and VA-146 *Blue Diamonds* as well as Douglas F-4B Phantoms of VF-142 *Ghostriders* and VF-143 *Pukin Dogs*. Their targets were torpedo boat bases at Hon Gai and Loc Chao. Another sortie of aircraft from *Constellation* attacked VPN boats that were at sea near Hon Me Island. Two vessels attacked by VA-145 were left badly damaged and on fire. During the attack, A-1H (BuNo. 139760) pilot Lieutenant (junior grade) Richard Sather of VA-145 call sign "Electra" was shot down by AAA offshore from Thanh Hoa. Thus, Sather became the first Navy KIA by hostile fire in the Vietnam conflict.

As the Skyhawks of VA-144 dived on their targets at Hon Gai AAA barrages filled the sky, with the ferocity taking the attack force by surprise.[31] Endless bursts of flak and glowing phosphorus tracers streaked past their aircraft, leaving clusters of black blotches in an otherwise blue sky. The skilled pilots had to constantly maneuver their aircraft, and jink from side to side, making AAA acquisition more difficult. Spotting four torpedo boats tied up on the west side of the bay, A-4 (BuNo. 149578) pilot Lieutenant (junior grade) Everett "Alvy" Alvarez of VA-144 call sign "War Paint 411" fired off his salvo of Zuni rockets on the first past, setting the boats ablaze. Pulling out and circling back for another pass, he opened up on a larger vessel in the bay with 20mm cannon fire. Spotting his flight leader flying low and in a southerly direction out to sea, Alvarez began maneuvering to join up.

As he approached a small isthmus of land he was hit by AAA, violently jarring his aircraft. As the Skyhawk began to roll to the left, and with smoke filling the cockpit, Alvarez ejected low to the water. After splashing down in Ha Long Bay an orbiting Skyraider tried to protect him from an approaching flotilla of North Vietnamese boats. Despite this effort, he was quickly captured. Alvarez has the unfortunate distinction of becoming the longest-held Navy POW of the Vietnam conflict, spending eight and a half years in captivity.

At the time of *Pierce Arrow* in August 1964, the United States had 23,300 military personnel in theatre. By 1965 that number had ballooned nearly 800% to 184,300, and would more than double to 385,300 in 1966 and grow to 485,600 in 1967. The bulk of American air losses before *Pierce Arrow*[32] were sustained by the Air Force and CIA in operations such as *Farm Gate*, wherein WWII and Korea-era aircraft were used by both American and VNAF forces to fight communist insurgents. However, the events of August 1964 would formalize the conflict, opening North Vietnam to attack, and bringing a fettered might of American air power to bear on the country.

On 10 August the Congress of the United States passed the Gulf of Tonkin Resolution, greatly expanding American military involvement in Indochina. The following year, on 2 March 1965, *Operation Rolling Thunder* would begin, the largest sustained military aerial campaign since WWII. Six days after the commencement of *Operation Rolling Thunder*, on 8 March 1965, 3,500 Marines of the 9th Marine Infantry Brigade arrived in Da Nang, the first American combat troops in the country. Two months later, in early May, the aircraft carrier *Constellation* arrived to Dixie Station in the South China Sea and began attacks on targets in South Vietnam as well as *Operation Steel Tiger* missions. This operation targeted an area of southeastern Laos where the Ho Chi Minh Trail infiltrated both Vietnams via the Mu Gia and Ban Karai passes. These were primary infiltration points for the VC into South Vietnam. Thus they would be subject to countless attacks during the conflict.

For the United States, what had previously been largely an advisory role for the military had become, in less than one year, a full-blown regional conflict. This would result in global ramifications that would have profound effects on America, and the world, for decades to come. It was thus a very different environment when on 19 October 1965 *Kitty Hawk* and CVW-11 departed NAS North Island in San Diego for their second WESTPAC cruise to Vietnam. While their first cruise in 1963-1964 had not been a combat deployment, their second certainly would be. It was during this subsequent cruise that VA-113 suffered its first fatalities of the conflict. On 17 December 1965 Lieutenant David Wickham, flying BuNo. 148510, was returning with wingman Lieutenant Bill Bowes from an armed *Steel Tiger* reconnaissance mission over Laos when they encountered bad weather and turbulence on a night approach to *Kitty Hawk*. Bill remembers the incident well:

> I was Dave Wickham's wingman on my 23rd combat mission which was a night road recce mission over Laos. We broke out and dropped bombs on the Mu Gia Pass, then joined back up and proceeded to the ship. We each went into our separate holding patterns until our respective push-over times and Dave was right ahead of me. I broke out of the clouds during the descent, and when I leveled off at 1,200 feet on the final leg of the approach I saw a large explosion in front of me. It was right on the ramp of *Kitty Hawk*. It was Dave's airplane, and he had a big ramp strike and was probably killed immediately.

The ship diverted the four airplanes that were still airborne to Da Nang. The next morning after the flight deck and debris were cleared we were called back. It was a very sad event I shall never forget.

On 11 April 1966 the CO of VA-113, Commander Henry "Hank" Dibble, rotated out. He was replaced by 37-year-old XO Commander John Abbott, a veteran of the Korean conflict who flew an F-4U Corsair with VF-53 from USS *Essex* (CV-9) from August 1951-March 1952. On 22 January 1952 Abbott, flying BuNo. 63033, and wingman Lieutenant Ed Laney, flying BuNo. 62943, were flying escort for four Skyraiders of VF-54 *Hell's Angels* over the coastal city of Chongjin, North Korea. Their targets were railroad infrastructure as well as a rail bridge. For unknown reasons, the engine in Abbott's Corsair quit and the cockpit began filling with smoke. He was able to glide the airplane to the coastline, where he bailed out. Though Abbott was able to extract himself from the stricken airplane and parachute into the water, he was still in great danger. Water temperatures in the Sea of Japan that time of year were near-freezing. In addition, Abbott landed in an area of the coast filled with maritime mines. Racing against time, an HO3S rescue helicopter was immediately launched from the cruiser USS *Rochester* (CA-124). Although ultimately successful, the extraction nearly cost the lives of both the rescue swimmer and wingman Lieutenant Laney, who was forced to bail out of his Corsair when it was damaged during the rescue operation.[33]

On 17 April 1966 VA-113 lost their second aircraft of the cruise, though thankfully with no loss of life. Lieutenant Bud Johnson launched from *Kitty Hawk* for a mission against the Hai Duong railroad and highway bridge thirty miles east of Hanoi. Immediately following the cat shot a suspected incorrect horizontal stabilizer trim setting caused Johnson's A-4 (BuNo. 148583) to pitch nose-down. He ejected and even though he landed in the water directly in front of the bow of the carrier, he managed to avoid being run over by or sucked under the massive vessel. While bobbing in the water following his ejection he noticed the expensive Breitling watch he had recently purchased while on liberty in Hong Kong was no longer working. A CH-53 Sea Stallion helicopter retrieved Johnson and his broken Swiss watch from the water.

Johnson credits Skipper Dibble for his survival, who always stressed to visualize and mentally note what actions to take should a cat shot go bad. He began to practice this religiously under Dibble, and on that day was able to recognize he had a problem about halfway through the shot. When the A-4 left the flight deck and pitched nose-down, Johnson was prepared.

On 20 April, just nine days after taking command of VA-113, Commander Abbott was shot down over North Vietnam. He was leading a group of Skyhawks on a flak suppression mission to the Vinh Son Highway Bridge over the Song Ca River, 25 miles northwest of Vinh. Abbott's aircraft (BuNo. 148512) call sign "Battle Cry 314" was hit in the starboard wing by AAA, resulting in a fire. VA-113 pilot Lieutenant Paul Adams was one of four Skyhawks on the mission that day, which he clearly remembers:

> While on my attack run against a flak emplacement I heard a voice on the radio say "I'm hit". On climbing out from my run I looked at my nine o'clock where Skipper Abbott and his wingman Lieutenant (junior grade) Harry Welch were. I saw streaks of fire on a dive to their target.

148

I completed my run and headed east while climbing and looking north where the skipper and Harry would be, and spotted a smoke trail. I declared an emergency on guard frequency and followed the smoke until it arced over and went down to the ground. Not observing an impact, I started a left circle to the north around the smoke trail, saw a good parachute at about my altitude, and watched until it touched down in a clear, grassy area near some trees. On my second circle Harry, also orbiting over the skipper, appeared in my field of view slightly lower and banking inside my radius of turn.

Welch also recalls that mission:

After Skipper Abbot ejected I made numerous passes over him. I saw him on the ground but never saw any sign that he was conscious. I knew that Paul had called for rescue, and CSAR helos were inbound. I made my last pass low, at about 100 feet, in the hopes of seeing some sign from Abbott, something that could be of some assistance in the rescue attempt. That is when I was hit.

While flying low and trying to keep track of Abbott on the ground Welch was also hit by AAA. His A-4 (BuNo. 149495) instantly became a fireball. As Adams describes it, Harry's Skyhawk appeared "Like a flaming comet with the nose of an A-4 sticking out".

VA-113 *Stingers* 1965-1966 *Kitty Hawk* WESTPAC
Front Row: Rubottom, Daniels, Willett, Abbott, Dibble, Scott, Lassey, Bronson, Murphy Back Row: Rypel, Hughbanks, Schubert, Churnez, Naughton, Utter, Grabarino, Graham, Adams, Bowes, Johnson, Greenamyer, Ellis

Commander Hank Dibble

Commander John Abbott

Commander James Burnett with
Skipper Holloway.

"Wild Bill" Ellis

Adams escorted Welch to the coast to make sure he made it feet wet on his way back to *Kitty Hawk*. At one point he advised Welch to eject from his burning aircraft before fire reached the controls or fuel tanks. His short and terse reply was "F**k you, it's flying". He managed to get his badly damaged aircraft 80 miles back to *Kitty Hawk*, where he was forced to eject after fire finally reached the aircraft controls. Welch was picked up by an SH-2 Seasprite plane guard helicopter from the carrier.

After Adams made sure that Welch was safely on his way back to *Kitty Hawk* he turned back to the area where Abbott had been shot down. While en route he called his wingman Lieutenant Andy Churney, learning he had egressed earlier from the area due to a suspected bird strike, leaving Adams as the only A-4 remaining.

When he arrived back over where he had last seen the unconscious Abbott the skipper and his parachute were no longer visible. As Abbott was carrying a rescue radio Adams made several attempts to contact him, however, there was no response. After relaying a situational and position report to rescue forces, he requested a tanker to refuel and stay on station to assist. When he informed the carrier there was no sight of Abbott he was ordered to return to *Kitty Hawk*.

Abbott disappeared without any communication. After the conflict ended it was discovered he died seven days after being shot down. Whether this was from injuries suffered during the ejection, at the hands of his captors, or a combination thereof remains unknown. After the loss of Abbott, it was Lieutenant Commander Richard "Dick" Willett who informally took over leadership of the squadron until the end of April when Commander James "Andy" Burnett came aboard and formally assumed command while *Kitty Hawk* was on a break from the line at Subic Bay in the Philippines.

At the end of May, during the next and last break from the line period at Subic, Commander Bob Bennett came aboard and took command of the squadron. Commander Burnett then assumed the role of XO. Bennett was also a veteran of the Korea conflict, where he had been a member of VA-195 *Dambusters* flying the AD-4 (A-1) Skyraider. On 1 May 1951, flying from USS *Princeton* (CV-37), Ensign Bennett participated in a historic mission to drop aerial torpedoes on the strategic and heavily defended Hwacheon Dam in North Korea. Though many previous attempts had been made by both Air Force and Navy squadrons, it was VA-195 that finally accomplished the task, which earned the squadron the nickname *Dambusters*.

The skipper had also previously served a 1960-1961 tour with VA-113 aboard USS *Hancock* (CVA-19), where one of his squadron mates was future astronaut Eugene "Gene" Cernan, the last person to set foot on the moon. In an act of humble modesty, Bennett never told the squadron of his participation in that famous mission in Korea, nor of having previously served in VA-113 with future astronaut Cernan. It was not until decades after they served together that the men he commanded fully discovered his pedigreed past.

The 1965-1966 WESTPAC ended just weeks after Bob Bennett took command. The tour had been costly for the squadron, having lost two pilots and four aircraft. However for *Kitty Hawk* and CVW-11 as a whole, over the course of eight months 25 fixed-wing aircraft were lost. The cost in lives was 19 men who would never again return. Over half the losses came in the last two months of the cruise, during attacks on heavily defended targets of industry and infrastructure.

Though the loss of two lives had been painful for VA-113, they were not in vain. The *Stingers* of that cruise agree the valuable experience gained is what molded them into an efficient combat squadron. Their skills had been honed under fire; they had become a hardened and effective fighting team. These skills would be further honed and passed down throughout future deployments to the theatre. Equally critical as the development of combat airmanship skills were the personal bonds that were developed; a close kinship that can only be understood by those who have fought and served in combat together. These bonds were a key part of their motivation to serve and fight not only for their country but for each other as well. As *Kitty Hawk* began the long trek across the vast Pacific Ocean to its homeport of NAS North Island in San Diego, the minds of many in VA-113 were thinking forward five months, to November. They would then become part of CVW-9 embarked on *Enterprise*, and once again fly into the increasingly dangerous, and deadly, skies of North Vietnam and Laos.

As was customary for combat squadrons of the time, during their five-month turnaround period at NAS Lemoore VA-113 saw the departure of about half its pilots, along with an equal influx of fresh faces. While five of the new arrivals were experienced naval aviators, some with previous combat cruises, seven of the new pilots were "nuggets"—fresh from the RAG on their first combat deployments. Nuggets arriving from the RAG had minimal combat training, all in practice and theory. The actual skills and knowledge needed to fly, and survive, in the Vietnam theatre could not be taught in a classroom or in a controlled learning environment that lacked enemy opposition. They could only be learned by enemy fire while under the watchful guidance and instruction of combat veterans.

"Heinemann's Hotrod" with its distinct
delta wing and cruciform empennage.

"Six seconds of do or die, or both"
Mk.82 bombs on their way to target.

Among the veterans remaining for the subsequent cruise was Lieutenant Commander Ted "Cash" Bronson. He acquired his nickname when, after a mission against a fortified bridge during the 1965-1966 *Kitty Hawk* cruise, he strode into IOIC claiming several direct hits and boldly wagered that if the bridge was still standing he would forfeit one month of pay. The BDA photographs came in, which showed the bridge still standing. It was a wager the rest of the squadron would never let Cash Bronson live down.

Lieutenant William "Bill" Bowes had joined up with VA-113 in June 1965 shortly before the squadron deployed aboard *Kitty Hawk* for their second WESTPAC tour. A Navy ROTC alumni, he earned his wings in December 1964 and reported to VA-113 at NAS Lemoore six months later.

Lieutenant Commander Tom "Hook" Scott was another veteran of the previous *Kitty Hawk* cruise. A 1956 graduate of USNA, by the time he joined VA-113 in January of 1966 he had already served one A-4 combat tour in Vietnam with VA-195 on *Bon Homme Richard*. He later became known as "One-Shot Scott" after he accidentally discharged his .38 revolver sidearm into the deck of the squadron ready room.

Lieutenant Bill "Wild Bill" Ellis holds a very special place in the hearts of those he served with in VA-113. He had just turned 18 when he enlisted in the Navy in November 1962. By July of the following year he had graduated from "A" school as an electronics technician. Quickly advancing to E-4, in March 1964 he took the entry test for the NAVCAD program. His score was so high he was immediately selected for AOCS and flight training. On 1 February 1965, Bill received his commission and wings, five months before his 21st birthday.

That an eighteen-year-old with no college or flight experience could go from non-rated enlisted to commissioned officer and jet combat pilot in just over two years speaks volumes to Bill's intelligence and ambition. He reported to VA-113 on *Kitty Hawk* in November 1965 while the carrier was in Yokosuka, Japan. The squadron didn't quite know what to do with the young pilot who, aside from the gold wings on his uniform, did not fit the stereotypical portrait of a naval combat aviator. Instead of the "Type A" assertive and competitive personality traits found among many of his peers, Bill was humble and unassuming. However, it was not long before he showed himself to be an aggressive and skilled combat pilot, earning the trust, respect, and admiration of his squadron mates. Paul Adams recalls:

> Bill was my wingman on his first night combat mission over North Vietnam. In pre-briefing for the mission I painstakingly went through everything in great detail, assuming I needed to hold his hand because of his age and inexperience. As it turns out he performed magnificently, beyond all my expectations.

By the time of the 1966-1967 cruise on *Enterprise* Bill had become a combat section leader and an integral member of VA-113. Along with Wild Bill he was also affectionately known among his squadron peers as "Baby Stinger Bee".

The slew of nugget aviators also included Ensign Jay Greene, who had an interesting background and rather unexpected path to Naval Aviation. He had attended Menlo Junior College before transferring to Stanford. While there he was minutes away from taking the oath of enlistment for the Army Reserves when the recruitment officer, a local policeman, was called away to deal with a civilian law enforcement matter. After graduating college and getting married he worked several mundane jobs where the daily doldrum left him desiring more. The newlywed and his wife then joined the recently formed Peace Corps, with an eye on travel and adventure. While training in Puerto Rico fate intervened and instead of traveling to far-flung exotic destinations with his bride as part of the Peace Corps, Jay Greene found himself applying to the Air Force to become a pilot. A minor medical anomaly led them to cautiously disqualify him for pilot training, instead offering him non-flying positions. Being a long-time admirer of military aviators such as Duvall Hecht, his former English professor and crew coach at Menlo, who was also a gold medal olympian and former Marine Corps aviator, he turned down the offer. He instead applied to the Navy, who happily accepted him. He was soon off to AOCS in Pensacola, where the former Peace Corps volunteer would begin learning how to drop bombs and strafe targets.

Ensign Ernie Christensen was a 1964 graduate of USNA, earning his wings in 1966. Ernie's Danish grandfather Rasmus had emigrated to America from Germany, arriving in Florida in 1904. He later enlisted in the Navy and after a decade of service saved several men when a boiler exploded on the destroyer he was serving on. As a reward, he was sent to Pensacola, which at that time was becoming the birthplace of Naval Aviation. By 1915 he had been assigned to seaplane aviation as a master machinist. He eventually became an aviation machinist then attended Navy flight school in Pensacola, becoming naval aviator number 1885.

In 1919 Rasmus was a flight engineer on NC-1, one of three Navy Curtis seaplanes to attempt the first transatlantic crossing by air, eight years before Charles Lindbergh completed his solo flight. Originally four seaplanes were scheduled to attempt the crossing however NC-2 was cannibalized for parts before departure.

NC-1 was damaged after landing in the Atlantic Ocean during foul weather. Unable to become airborne again, the crew was rescued by the Greek cargo ship *Ionia*. The aircraft was then taken under tow, however, it sank in deep water three days later. NC-4 was the only aircraft to successfully make the crossing. Rasmus, along with the rest of the crew from NC-1, were reunited with the crew of NC-4 when it landed in Horta, Azores Islands.

One of the pilots of NC-1 was 32-year-old Lieutenant Commander Marc Mitscher,[34] a future admiral who would play a critical role in the Allied victory in the Pacific theatre of WWII. The family tradition of naval service continued with Ernie's Father Ernest Christensen Sr. who graduated from USNA in 1934 and earned his wings in 1937. During WWII he served in several capacities, including skipper of USS *Rehobeth* (AVP-50), a seaplane tender based in the Atlantic. Thus Ernest Christensen, Jr. became the first third-generation aviator in Navy history. In a twist of irony Captain Ernest Christensen Sr. would one day command the aircraft carrier USS *Hornet* (CV-12), the namesake of the first carrier *Hornet* (CV-8) once commanded by Captain Marc Mitscher, who had flown with his Father on NC-1. The senior Ernest Christensen retired as a rear admiral in 1971.

Ensign Thomas Patrick "Pat" Anderson had been a childhood friend and schoolmate of Ernie Christensen at Punahou School in Hawaii and Severn School in Maryland. His Father was Admiral George W. Anderson, Jr. who had served as CNO 1961-1963. Pat had earned an appointment to USNA, however, he felt being the son of an admiral would draw undue attention, so he opted for Duke University and Navy ROTC. He entered flight school in 1964 and went through the training pipeline with his old friend Ernie Christensen.

They graduated and earned their wings at the same time, then both were assigned to VA-113 just before the cruise on *Enterprise*, where they shared a stateroom. Of his old friend and parallel paths Ernie remembers "I have little doubt as I look back that Admiral Anderson had a hand in all of our similar orders at similar times and, it wasn't until we left VA-113 that we finally parted as roommates".

Ensign Don Williams had attended Purdue University on a Navy ROTC scholarship. An engineering major, he was initially not interested in aviation. A summer program at NAS Corpus Christi changed that when he had the opportunity to take the controls of an aircraft and pilot it. When the time came he applied for and was accepted into flight school. After primary training he applied for jets, and with his high academic and airmanship scores, he was soon learning to fly the A-4.

Another new arrival to VA-113 was Lieutenant Commander Jeremy "Bear" Taylor. However, Bear was far from a nugget. He was a highly experienced naval aviator, and a veteran of Caribbean, Mediterranean, and North Atlantic cruises aboard the USS *Franklin Delano Roosevelt* (CVA-42) and USS *Shangri-La* (CVA-38). Bear had entered the Navy as a NAVCAD in 1955. A boisterous and magnetic personality, as well as a natural leader and bold and uncompromising warrior, Taylor would go on to play a key role in VA-113 history.

Jay Greene 1966-1967 WESTPAC on *Enterprise.*

Bear Taylor 1968 WESTPAC on *Enterprise.*

On 22 November 1966, three days after they departed from San Francisco, the bow of *Enterprise* broke the blue waters of Pearl Harbor. For the next six days, the air wing and ship's company underwent ORI, ensuring they were prepared for combat. VA-113 flew numerous day and night training missions during the ORI period, keeping their skills sharp as their combat deployment on Yankee Station loomed. Many in the squadron spent their last evening in Honolulu enjoying a good meal. *Enterprise* took up lines and departed Pearl for Subic Bay on 28 November.

On 2 December *Enterprise* was 2,100 miles west of Pearl Harbor when a Soviet Tupolev Tu-95 Bear bomber flew over the ship. An F-4 Phantom from the carrier intercepted the bomber and flew escort as it passed through the area. Another Tu-95 would make a similar appearance two days later, though on that occasion it refrained from flying directly over *Enterprise*. The appearance was less a show of force from the Soviets than it was an opportunity to gather intelligence, which was assuredly passed onto the North Vietnam.

During much of their time in transit to Subic Bay, the only flying done by VA-113 was in the classroom. Daily lectures on topics such as survival techniques, SAR operations, squadron combat tactics, and SAMs were part of the curriculum, along with classes on the politics, policies, and current events within North and South Vietnam. By 7 December *Enterprise* was 220 miles northwest of Manila when the squadron began flying practice missions over the Philippines, reconnoitering roads and dropping bombs on stationary targets. The next afternoon, 8 December, the carrier tied up at Subic Bay, the last port of call before arriving on Yankee Station.

31 January 1967 VA-113 en route back to Yankee Station from Subic Bay

Front Row: Wales, Christensen, Taylor, Williams, Lassey, Burnett, Bennett, Greenamyer, Bowes, Holton, Ellis, Bronson
Back Row: Deeter, Arnold, Erie, McGraw, Lenhard, Brown, Brennock, Naughton, Graham, Greene, Anderson, Scott, Johnson, Phillips

There were additional lectures at Subic Bay, including one from CVW-2 CAG Commander William Harris from *Coral Sea*, who related the experiences of the air wing flying combat in Vietnam. CVW-2 suffered high losses during their second WESTPAC cruise, which commenced on 29 July. By the time Harris arrived aboard *Enterprise* on 9 December to present his lecture *Coral Sea* had lost 14 aircraft in the Vietnam theatre in just over four months.

Another lecture taught the use of the AN/ALQ-51 ECM units newly installed in VA-113 A-4Cs. ECM was designed to electronically disrupt the radar tracking capabilities of North Vietnamese Fan Song, Flap Wheel and Fire Can radar. While still in its infancy, the widespread use of ECM in the Vietnam theatre resulted in the rapid maturation of the technology. In later years of the conflict improved ECM resulted in a tangible difference in fewer aircraft lost to radar-guided AAA and SAM missiles.

On 12 December VA-113 underwent JEST training. Taught by Negrito tribesmen, the purpose of JEST was to prepare naval aviators for survival and evasion in the jungle should they be shot down. The Negrito taught how to use bamboo, where to find drinkable water, where to find cover and sleep, and how to make fire. The aviators were shown how to survive by eating insects, reptiles, and other creatures found in the jungle. Of his day spent in JEST training Jay Greene wrote in his diary:

> During the day a Negrito caught a fruit bat. It was about three feet from wingtip to wingtip, had a hairy body, and a face like a dog. These bats were all over the place and looked like hawks soaring in the sky. We ate the bat plus some mudfish, crawfish, and rice for dinner.

On 15 December *Enterprise* departed Subic Bay for Yankee Station. Two days later the first combat flight schedules were posted in ready rooms. The time had come, the hour was at hand. On Sunday 18 December the air wing began flying their first combat missions. On that day XO Andy Burnett and wingman Jay Greene flew a hop to the Thuan Le POL site near Ha Tinh. It was Greene's first combat mission, under the watchful eye of XO Burnett. They were pleasantly surprised when they encountered no resistance.

The next day Jay Greene, Ted Bronson, and Jim Graham flew a mission where weather and poor visibility precluded the original target. Instead, they dropped their loads of Mk.82 bombs on a dirt-filled bridge near Ha Tinh. Again, there was no observed AAA. The foul weather continued on 20 December, when a mission to Vinh was scrubbed. A diversion to Laos found no targets. The following day saw a slight improvement in the *Steel Tiger* zone in the Southeast portion of Laos, where the Ho Chi Minh Trail winds in and out of both countries.

Bad weather continued over much of North Vietnam, leading to the cancellation of numerous missions. However, the weather to the east was better, allowing for more *Steel Tiger* missions over southeast Laos. On Saturday 24 December, Christmas Eve, a FAC mission to *Steel Tiger* concluded with the FAC wishing the flight of four VA-113 Skyhawks a Merry Christmas.

Christmas Day brought a truce in hostilities, with VA-113 celebrating in their ready room with music, cocktails, and the showing of home movies. The respite was short-lived, however, for in the darkness of the next morning combat operations resumed. Inclement weather continued to hamper efforts until early January when it improved enough for missions to the cities of Haiphong, Thanh Hoa, and Vinh.

Skipper Bennett, XO Burnett, Greene, Bronson, Lenhard, Lassey, and Greenamyer flew multiple hops, attacking a radar station and railroad cars. They also hit several coastal barges with rockets, bombs, and guns.

Christmas 1966 in VA-113 ready room.

On 12 January VA-113, as part of a coordinated strike group, flew missions against two bridges thirty miles southwest of Thanh Hoa. Among the eight A-4s in the strike group, two were flying as anti-SAM "Iron Hand" aircraft, with AGM-45 Shrike missiles at the ready. Also in the strike group were two A-6s and two F-4s attacking with AGM-12 Bullpup missiles as well as Mk.82 and Mk.83 bombs. The strike failed to inflict any serious damage to the bridges. This was not unusual, as many North Vietnamese bridges targeted by American forces had an uncanny ability to withstand even the fiercest of attacks. This was a scenario that would be encountered numerous times over the course of the conflict—attacks on bridges that failed to inflict irreparable damage.

The Bullpup was a rocket-propelled, precision-guided missile, first used in the Vietnam theatre. It was guided by the pilot with a directional button, which on the A-4C was located on the top of the control stick. To deliver the Bullpup the pilot would dive on the target using the "pipper" on the gun sight. Once lined up the pilot would "pickle" the missile away. The Bullpup would roar as its rocket engine ignited, which would momentarily obscure pilot visibility and cause the aircraft to buffet. A flare on the missile would assist the pilot in visually guiding it to the target. Though the Bullpup was an effective weapon, it also had several drawbacks. Guiding it took considerable skill from the pilot, for as it nears the target the missile is traveling close to Mach 2. The slightest overcorrection at that speed could easily cause the missile to miss the target. In addition, the Bullpup was not a "fire and forget" weapon. Because the pilot had to monitor and guide the missile from the time of firing to the time of impact, they had to maintain visual contact with the weapon. This means they were limited in maneuverability during that time, making them highly vulnerable to AAA, SAMs, and enemy MiGs.

CVW-9 had two Skyhawk squadrons, the other being VA-56. Two days after the bridge strike, on 14 January, two A-4s from VA-56 collided mid-air over Laos during a night combat mission, with both pilots successfully ejecting. The next morning, as Lieutenant Commander Arthur Tyszkiewicz call sign "Champion 411" (BuNo. 145087) was being hoisted aboard a USAF rescue helicopter, an equipment malfunction sent him plunging back to the jungle floor. This most likely ended his life, as the rescue crew was unable to reestablish visual or radio contact with him. The second pilot, call sign "Champion 407" (BuNo. 147724), was rescued that same morning by a USAF Huey UH-1F helicopter. That pilot turned in his wings two weeks later.

In May 1968 the remains of Tyszkiewicz were discovered by American forces in the jungles of Laos and repatriated. On 18 July of that year, off the coast of California, his final request was honored by the Navy. His remains were buried at sea from *Enterprise*, four hours before the carrier returned to NAS Alameda, ending the 1968 WESTPAC. The loss of Tyszkiewicz was strongly felt in VA-113, where he had several close friends such as Bear Taylor. Having previously served together, he had been best man at Taylor's wedding.

The last mission of the line period occurred on 16 January with a coordinated strike against the Than Linh Dong Bridge, located fifteen miles northwest of Hanoi. That evening *Enterprise* turned east and steamed toward the Philippines, where the crew would enjoy a much-deserved break, arriving in Subic Bay at 1800 on 18 January. Many officers not on duty immediately headed to an O-Club for dinner, followed by a visit to one of the many lively and unusual nightlife spots in nearby Olongapo. A popular item to buy in the Philippines was furniture and decorative accents made of wood from native tropical Monkey Pod trees. However, in an example of *caveat emptor*, many purchasers found out the hard way that the wood had not been properly prepared. Within a few months of arrival in the continental United States, the items would begin cracking and splintering.

On 22 January the squadron held a party on Grande Island, playing softball and indulging in BBQ steaks and beer. The eight-day visit to Subic Bay was followed by three days in Manila.

At 0800 on 30 January *Enterprise* departed Manila and spent the remainder of the day practicing fighter intercepts with aircraft from the British aircraft carrier HMS *Victorious* (R38). The following days inclement weather once again prevented any flying. This gave the *Stingers* an opportunity to head topside for a squadron picture.

On 4 February the weather cleared enough for a massive Alpha strike against the Thanh Hoa railway siding involving 29 aircraft from *Enterprise*, *Ticonderoga*, and *Kitty Hawk*. The target was well-defended by thirty-five 37/57mm AAA sites and ten radar-controlled 85mm AAA sites. In addition, the railway siding was well within the range of fourteen confirmed SAM sites.

Bill Bowes was assigned flak suppression in an area containing multiple revetted 37mm AAA guns. Approaching the target Bowes climbed to 15,000 feet, which would give him plenty of altitude to guide a Bullpup missile into the revetments. Firing the missile, he stayed focused on the target despite heavy and accurate AAA fire enveloping his aircraft, guiding it into a revetment. A tremendous fireball erupted, with two large secondary explosions, silencing the gun permanently. For his actions that day Bowes was awarded a DFC. He clearly remembers:

I easily saw the target on what I recall was near the southeast side of the Thanh Hoa Bridge. I rolled into a 30° dive and released the Bullpup. The rocket motor lit, and I immediately saw the flare that is on the back end of the missile, used for visual guidance by the pilot. Flak started coming up and black puffs were all around.

I just concentrated on the Bullpup and guided it to the AAA site which was firing at us. They well knew I was aiming the missile at their emplacement. I stayed focused on guiding the missile until impact and then immediately broke left, climbing at full power to get above 3,000 feet, jinking left and right, hoping to avoid getting hit. I did not see the secondary explosions that were later verified, but I knew this missile had hit its target.

Extensive AAA damage to wing.

AAA over Haiphong.

The next day Jay Greene and Bud Johnson attacked the causeway linking Hon Nghi Son ("Cigar") Island with the mainland, using 250 lb. Mk.81 bombs, also known as "ladyfingers" or "firecrackers," to accomplish the task. Several *Steel Tiger* bombing missions to Laos took place over the following days. This included a "Blind Bat" mission wherein a C-130 or other aircraft would drop illuminating flares over an area, making the targets visible in the darkness of night for attacking aircraft.

Intermittent foul weather continued to dog air operations up until 8 February, the first day of the Tet ceasefire. The ceasefire ended on 14 February, Valentine's Day. However, the weather remained a factor, allowing only limited recce flights and *Steel Tiger* missions. A break appeared in the weather on 16 February, allowing an opportunity to launch a large Alpha strike on the port city of Hon Gai, northeast of Haiphong. With the strike group only 20 miles from their target, the mission was scrubbed due to low cloud cover. This resulted in tons of costly ordnance being harmlessly dropped into the water.

On 20 February Taylor, Greene, Thomas, and Naughton flew a mission to the Ha Tinh rail bridge, some 25 miles south-southeast of Vinh near Route 1A. While Thomas and Naughton were each carrying five Mk.82 bombs, Taylor and Greene were each carrying two precision-guided AGM-12B Bullpups. Thomas and Naughton were dispatched by flight leader Taylor to hit the bridge first with their Mk.82 bombs. As they detached a call came on the radio from F-4 pilot Lieutenant (junior grade) Charlie Hill and his RIO Lieutenant (junior grade) André Frank of VF-96 from *Enterprise*. The two men, along with flight leader Major Russell Goodman and RIO Ensign Gary Thornton, were attacking railway facilities eight miles southwest of Thanh Hoa.

Hill first reported heavy flak over the target, followed by an announcement that Goodman and Thornton call sign "Showtime 614" had been hit by AAA. Taylor inquired if they could be of any assistance, and Hill answered in the affirmative. After flying some 100 miles north Taylor and Greene arrived at the scene only to be informed that Goodman and Thornton had ejected and there had been no visual or voice communication from them, therefore there would be no rescue effort. Now well north of their original target with Bullpups at the ready, Taylor radioed Greene that they were going after "*the* bridge". Thinking he meant the original target at Ha Tinh, Greene was surprised when Taylor instead took a heading to the infamous Dragon's Jaw Bridge at Thanh Hoa.

This vehicle and rail bridge was vitally important, as it was a critical transportation corridor from Hanoi and Haiphong to the southern areas of North Vietnam that abutted the DMZ and Ho Chi Minh Trail. It was one of the most strategically and ideologically important targets in North Vietnam, heavily defended by plentiful AAA and SAMs. Therefore, it was usually the subject of an Alpha strike with dedicated aircraft for flak and SAM suppression. Two light attack aircraft going after the bridge alone with only Bullpup missiles was a very risky endeavor. However flight leader Taylor decided it was worthy of an attack, a veritable A-4 "Charge of the Light Attack Brigade". Two men against one of the most heavily defended targets in North Vietnam.

Taylor's first missile hit the center of the bridge, while Greene's first missile malfunctioned and would not respond to guidance commands. Lining up for a second run the intensity of numerous flak sites around the bridge filled the sky with a barrage of deadly AAA fire. Greene remembers:

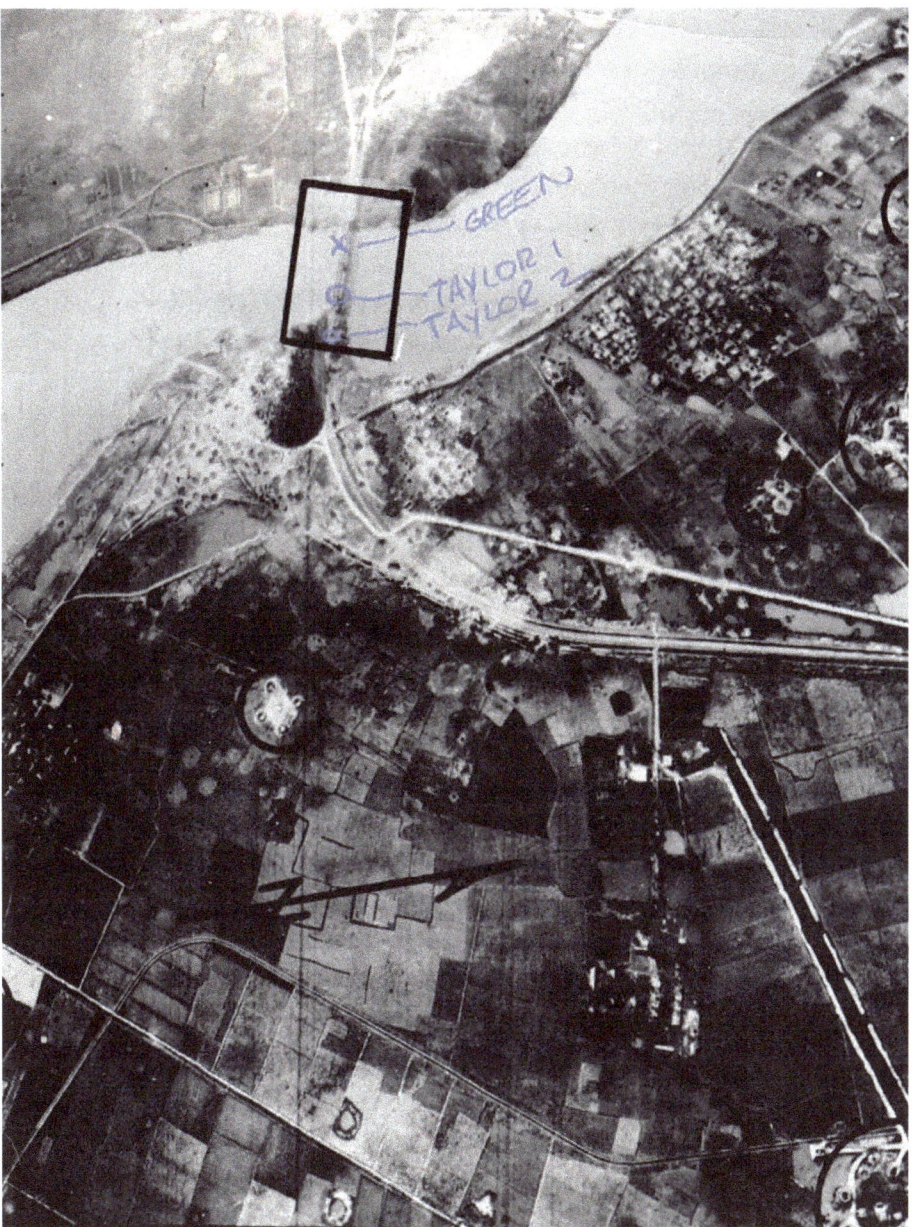

Post-BDA photo of Dragon's Jaw following the attack by Taylor and Greene 20 February 1967. Air defenses are circled in black.

> Flak sites around the bridge opened up. Fireballs were streaming by the canopy and I could see and hear the flak bursts around me. It sounded like someone was snapping their fingers on the skin of the plane. I was sure I was getting hit.

Releasing his second Bullpup, Greene held steady on the target as long as he could guide the missile to the bridge. The AAA became so intense that, at the last second, he pulled the stick of his A-4 hard to the right, witnessing the missile impact the bridge on the western end. Taylor's second missile missed the bridge, hitting railroad tracks on the eastern end. With their missiles expended, they returned to *Enterprise*, where Taylor had to explain his decision to attack the bridge to an angry Skipper Bennett. To Taylor, there was no question about commencing an attack on Dragon's Jaw.[35] To not attack was contrary to the mission and purpose. He succinctly states his reasoning:

> Attack pilots worth their salt are imbued with the "spirit of attack born and carried in a brave heart". In my view, an attack pilot must choose the audacious course and attack. That is not vainglory. That is "fortune favors the bold." That is the "spirit of attack".

Like many other bridges in North Vietnam such as the Paul Doumer Bridge over the Red River that connects Hanoi and Haiphong, Dragon's Jaw would remain defiantly standing throughout numerous large-scale attacks, becoming a propaganda symbol of North Vietnam. It was not until *Linebacker* operations in April and May 1972 that the bridge was rendered inoperable by Air Force Phantoms. It was finally destroyed in October of that year by Navy A-7 Corsair II[36] attack bombers.

It was in late February when VA-113 AIO Lieutenant Gary Arnold began hinting that changes were coming in the air war. The first sign came on 26 February when six A-6 Intruders of VA-35 from *Enterprise* dropped mines in the Song Ca and Son Giang Rivers, a move senior military advisors such as future CNO Admiral Thomas Moorer had been advocating for years. The mission, led by VA-35 Skipper Commander Arthur Barie, marked the first time maritime mines were dropped from jet aircraft, an act that would be repeated many additional times over the course of the conflict.

In March the targeting list began expanding to include new objectives near and in Hanoi and Haiphong, which at the time were two of the most heavily defended cities in the world. VA-113 and other A-4 squadrons operating on Yankee Station were tasked with these dangerous missions. On *Bon Homme Richard* Commander Homer Smith of VA-212, the first squadron equipped with the Walleye, had led the first smart bomb attack against North Vietnamese Army barracks in Sam Son on 11 March. As with the Bullpup missile, the precision-guided Walleye was also used for the first time in Vietnam. It had a clear, domed nose containing a television lens mounted on a gimbal for stability. In the A-4 cockpit, the radar screen was removed and replaced with a miniature television screen that displayed the image from the camera in the nose of the bomb. Similar to the Bullpup, the pilot would dive on the target, keeping it centered in his gunsight pipper. Then, using the television display, the pilot would maneuver the aircraft until the crosshairs on the screen were centered on the target.

Squeezing the trigger on the control stick would lock the camera onto a contrasting image of light and dark, with a window or door of a building providing optimum contrast. As long as the pilot continued to depress the trigger the camera would maintain lock, with the gimbal keeping focus on the selected target even as the aircraft maneuvered. When the pilot determined that the Walleye was within glide range he pickled the weapon with his thumb. The bomb would drop and the television screen would go black. A "fire and forget" weapon, after release the pilot was free to maneuver away and escape before impact. The avionics of the Walleye were powered during descent to the target by a wind-driven generator in the tail. While the Walleye was not well-suited for hard targets, it was ideal for soft targets such as unreinforced buildings.

*Enterprise* left Yankee Station on 3 March, for a much-deserved two-week break from the line. Arriving in Subic Bay late the following afternoon, the piping of liberty gave way to an onslaught of men vacating the ship, eager for indulgences after more than a month at sea. Following a week in Subic Bay *Enterprise* took up lines and steamed to Hong Kong, arriving on 14 March.

At the time Hong Kong was a popular port of call for American sailors, a bustling and vibrant metropolis of the British Crown, teaming with life, activity, and bargains. Many in the air wing spent their liberty exploring, eating, drinking, and shopping, particularly for cameras and stereo equipment. On 20 March the carrier departed Hong Kong, setting a course for Yankee Station.

During the two weeks *Enterprise* and CVW-9 were off the line the Navy suffered 13 aircraft losses resulting in 1 POW and 34 fatalities. The worse of these losses occurred on 10 March when a Navy VC-47J Skytrain (BuNo. 99844) carrying administrative personnel from Cam

Ranh Bay to Saigon crashed near Phan Rang. Four crew and twenty-one passengers lost their lives in the tragedy. The cause of the loss remains unknown.

On the morning of 22 March *Enterprise* resumed air operations. However, bad weather once again reared its ugly head. It was not until 25 March that weather cleared enough for Jay Greene, George Wales, and Bud Johnson to bomb a bridge near Quang Phong. Once near the target rain began pouring, causing some of their bombs to miss the mark due to reduced visibility.

On Sunday 26 March, while millions of Americans were celebrating Easter, it was business as usual on Yankee Station. Jay Greene, Tom Scott, Bob Naughton, and Bob Brennock returned to the area of Quang Phong with five Mk.82 bombs each to attack a rail bridge. Greene was the only one who hit the bridge, at the southern end. Following their attack on the rail bridge, they strafed barges suspected of carrying munitions with 20mm cannon fire.

The following morning a planned Alpha strike against the Hon Gai TPP was aborted just before launch due to bad weather. Later that same day Jay Greene returned to the bridge north of Quang Phong with George Wales, Tom Holton, and Tom Brown, with all their Mk.82 bombs missing the target. However Brown scored a direct hit on the main bridge with a Bullpup missile, taking it out of service. Also on 27 March an attack by VA-112 *Broncos* from *Kitty Hawk* on a road bridge near Dao My, 20 miles northwest of Vinh, resulted in the loss of an A-4 (BuNo. 148519). Pilot Lieutenant Alexander "Doc" Palenscar, a 1963 graduate of USNA, remained MIA until November 2002 when his remains were positively identified.

On 30 March Jay Greene, Tom Scott, and Jerry Greenamyer flew with CAG Shipman on a recce mission of Route 15 near Ha Tinh, an area referred to by pilots as "Happy Valley". A flak site opened fire, however, their aim was far off the mark. Greene would repeat the recce mission with CAG Shipman on 1 April, with both aircraft strafing barges and a motorized junk.

On 2 April Greene and Bill Ellis flew a nighttime road recce mission of Route 1A south of Ha Tinh. The weather on this pitch-black night was the worst encountered on the cruise thus far, with fog and a 450-foot overcast ceiling, causing flight operations on *Enterprise* to be canceled. Despite the horrendous weather both Greene and Ellis landed on their first attempts. Of the landing, Greene later wrote:

> I went on to log more than 43 additional night carrier landings and never encountered another more challenging. In this way, so early in my carrier experience, I gained great confidence and was able to say, over and over, "I've landed under worse conditions; I can do this".

By 7 April the weather had improved, though only marginally. Four A-4s of VA-113 flew into Laos on a FAC-guided *Steel Tiger* mission. They put all their Mk.82 bombs on target, eliciting a compliment from the FAC on their accuracy. A second *Steel Tiger* mission later that day proved less successful, as an encroaching thunderstorm gave the aircraft time for only one run at the target. The next two days the weather was again marginal. Though an Alpha strike on the Haiphong TPP was canceled due to low clouds, conditions proved good enough for more recce missions along the coast, using rockets and guns on targets of opportunity.

Aircraft that were unable to trap on a carrier on Yankee Station due to weather, hung ordnance, fouled deck or other factors had the option to "bingo" to Da Nang Air Base in South Vietnam, a one-way distance of roughly 100 miles. On 10 April Jay Greene, Tom Scott, Tom Brown, and Bill Ellis attacked a truck convoy near Dong Hieu. However, Bill's 2.75" rockets would not fire from the LAU, necessitating a bingo flight to Da Nang, accompanied by Tom Brown, to remedy the problem.

On 10 April an A-4 call sign "Jury 212" (BuNo. 151200) of VA-192 *Golden Dragons* from *Ticonderoga* was damaged by AAA while on a recce mission at the mouth of the Lach Van River, 20 miles north of Vinh. This left pilot Lieutenant Commander Shattuck unable to control thrust, precluding in-air refueling. He was forced to eject and was picked up by a Navy SAR helicopter.

On 12 April XO Andy Burnett, Karl Lassey, Bill Ellis, and Jay Greene targeted a railroad bridge 20 miles southwest of Thanh Hoa. Thunderstorms made the attack difficult, with the aircraft having to penetrate a squall line to dive on their targets. Thunderstorms continued into the next day, with weather so bad that Pat Anderson and Bud Johnson had to divert to Da Nang.

The last combat missions of the line period were flown on 17 April. Skipper Bob Bennett, Ernie Christensen, Ted Bronson, and Jay Greene flew to Laos for another *Steel Tiger* mission. The *Stingers* easily rendered their target, a AAA site, inoperative. After going off the line and before arriving in Subic Bay *Enterprise* once again joined up with the British Navy carrier *Victorious* for weapons training exercises. The day was not without incident, however, as Skipper Bennet and Walt Lenhard flew through a vicious hail storm, inflicting damage on their Skyhawks.

*Enterprise* arrived at Subic Bay on the afternoon of 19 April. During their week in the Philippines air wing intelligence officers kept close track of the events on Yankee Station, and in North Vietnam. The pace and intensity of Alpha strikes had increased markedly. Missions to heavily defended areas of Hanoi and Haiphong were becoming commonplace.

On 24 April another attack on Kep Airfield took place, while 60 miles to the southeast a simultaneous attack was conducted against a railroad yard in the port city of Hon Gai. An F-8C Crusader (BuNo. 146915) of VF-24 from *Bon Homme Richard* was providing flak suppression for a flight of eight Skyhawks when it took a direct hit from 85mm AAA. The remains of pilot Lieutenant Commander Edwin Tucker were repatriated in 1987 and interred at Arlington National Cemetery the following year.

On 25 April, while the USAF conducted a raid on an electrical substation on the outskirts of Hanoi, 60 miles to the east the Navy pounded an ammunition depot and POL site in and near Haiphong. The attack was costly, with the loss of three Navy aircraft. An A-4C call sign "Sun Glass 663" (BuNo. 147799) of VA-76 *Spirits* from *Bon Homme Richard* was shot down by 23mm cannon fire from a MiG-17. Pilot Lieutenant Charles Stackhouse spent nearly six years in North Vietnamese captivity before being released during *Operation Homecoming.*

Lieutenant A.R. Crebo was flying an A-4E Skyhawk call sign "Flying Eagle 225" (BuNo. 151102) of VA-212, also from *Bonnie Dick*. He was at 8,000 feet and beginning his roll into the target when an SA-2 exploded near his aircraft. He had completed his run to the target when the Skyhawk became uncontrollable from the damage. Crebo managed to regain control of the Skyhawk and make it back to the carrier group, where he ejected and was rescued by a destroyer.

Also over Haiphong that day a flight of A-4Es of VA-192 from *Tico* were providing Iron Hand SAM suppression for attacking aircraft. Lieutenant Commander F.J. Almberg call sign "Jury 204" was several miles north of the city and had just fired a third Shrike missile at a SAM site when an SA-2 exploded close to his aircraft. With no discernible damage, he continued his attack when a second SA-2 detonated in close proximity, resulting in the loss of hydraulics. Almberg regained control of the aircraft and headed feet wet, ejecting 40 miles southeast of Haiphong. He was plucked out of the water by a Navy helicopter.

The following day, 26 April, two A-4s of VA-192 flying from *Ticonderoga* were lost on a mission to destroy a POL site near Haiphong. Lieutenant (junior grade) J. Cain call sign "Jury 200" was beginning his roll into a POL site when his aircraft was hit by AAA. He continued, dropping his bombs on the target and climbing away before smoke began filling the cockpit. He ejected from his crippled Skyhawk (BuNo. 152076), landing 15 miles south of Haiphong. Enemy coastal forces immediately began an effort to capture him but were repelled by a dozen aircraft composed of ten F-8 Crusaders and two A-1 Skyraiders. Cain was rescued by Navy helicopter twenty minutes after splashing down. The second A-4 lost from VA-192 that day resulted in the only Congressional Medal of Honor awarded to a Navy jet pilot during the conflict.

Lieutenant Commander Michael Estocin, who was operations officer of VA-192, had recently gained notoriety during a previous 20 April Iron Hand mission to Haiphong, destroying three SA-2 SAM launchers while taking heavy damage from AAA. Nursing his injured Skyhawk back to *Ticonderoga*, he landed in the barricade with his aircraft engulfed in flames.

On 26 April Estocin was once again providing Iron Hand SAM suppression for a mission against a POL site in Haiphong. He was leading a flight of three A-4s, flying in advance of the main strike force. After firing an AGM-45 Shrike missile at a SAM site he climbed to 12,000 feet to fire off another Shrike when an SA-2 was launched at his aircraft. Executing a last-minute maneuver to evade the missile, it detonated close to his Skyhawk, causing massive damage which left the A-4 aflame. The aircraft began a sharp descent, executing several rolls before coming back under some semblance of control. Turning feet wet Estocin began heading back to *Ticonderoga*, accompanied by an F-8 of VF-191 *Satan's Kittens.* The Crusader, piloted by Lieutenant Commander J.B. Nichols, attempted to make contact with Estocin via radio however there was no response. Twenty miles later Estocin's Skyhawk (BuNo. 151073) began rolling again, crashing inverted into the sea near the island of Dao Cat Ba, where he perished. Estocin was awarded the Medal of Honor posthumously.

On 27 April *Enterprise* departed Subic Bay and began the trek back to Yankee Station. The following day pilots attended a lecture about the events that had occurred during their break from the line. Over the course of nine days the Navy lost 8 aircraft resulting in 3 POWs and 2 fatalities. These losses reflected the increased focus on heavily defended targets of infrastructure in the Hanoi and Haiphong areas such as POL facilities, railroads, bridges, and power plants.

The first day back on the line, 29 April, saw another busy day for VA-113. Several missions were flown to an area south of Vinh, targeting a bridge and storage area. Skipper Bennett had trouble finding the bridge, resulting in misses for all four aircraft in the group. Bob Naughton ingested flak near Vinh, badly damaging his engine. He made it back to *Enterprise,* but the engine in his Skyhawk was a total loss.

Devastating attack on VPAF airfield resulting in a cratered runway.

Destruction of a hangar at VPAF airfield.

The day ended with the squadron receiving a briefing on the CIA-developed Fulton Recovery System, an aerial method of recovering human assets on the ground. The following day saw an attack on a POL facility south of Vinh, followed by strafing of barges near Ha Tinh. The day ended with a literal bang for VA-113 when a Bullpup missile detonated prematurely shortly after it was fired, badly rattling Karl Lassey and his Skyhawk, though with no damage to either.

The month of May began with an Alpha strike against a boat facility near Kep Airfield, which had been the target of multiple raids the preceding month. The strike against the boat facility was a diversion for another attack on Kep Airfield itself by aircraft from *Bonnie Dick*. The pilots remarked that there was little AAA and no SAMs over their target, but there were MiGs in the sky. No Navy aircraft were lost in the attack, nor the mission that took place later that evening against a POL facility 20 miles west of Vinh. Jay Greene remarked in his diary of the light 37mm and 57mm AAA fire encountered, and a rather rough night landing on *Enterprise* which blew all three tires on his Skyhawk.

On 2 May another Alpha strike was launched against NVA barracks at Chi Ni near Nam Dinh, the southern point of the Iron Triangle. Weather forecasts turned out to be inaccurate, as a low overcast cloud layer and thick haze resulted in only one mile of visibility. The strike was aborted, with VA-113 aircraft dropping their Mk.82 iron bombs on a storage site near Thanh Hoa. With better weather, the strike was again launched the next day with aircraft from *Enterprise* and *Hancock*. There was light AAA and no MiGs, however, the strike group encountered three SAMs near the target, one coming so close to an F-4 that the rocket motor burned paint off portions of the empennage.

The North Vietnamese commonly referred to American combat airmen as "air pirates" as they considered U.S. military action in the country unlawful and illegal. A reconnaissance photo of Vietnamese writing on hills south of Thanh Hoa that had caught the attention of airmen was translated by intelligence. It read:

THE ARMY AND THE PEOPLE ARE UNITED TO FIGHT AND WIN OVER THE AMERICAN AGGRESSORS. WE ARE DETERMINED TO ANNIHILATE THE AGGRESSIVE AMERICAN PIRATES.

VA-113 was only two months away from completing their combat tour when they suffered their first loss of the cruise. On 4 May 1967 Lieutenant (junior grade) Jim Graham call sign "Battle Cry 314" (BuNo. 148514) was flying as wingman to Ted Bronson, along with flight leader Lieutenant Commander George Wales and wingman Jay Greene. The target of the four A-4s was a SAM site northeast of Thanh Hoa, close to Sam Son and the coast. This area was a common ingress and egress route for Navy aircraft, thus air defenses were plentiful.

The target had already been attacked earlier that day, leading to assumptions that defenses would be minimal. Therefore operating procedures were altered. Instead of a 45° dive angle, the aircraft dove at a shallower 30°. The bomb release altitude was also changed from 5,000 feet to 3,500 feet. Though it was standard procedure to avoid making more than one attack run on defended targets, the briefing nevertheless called for a second pass. While these changes increased the odds of a successful strike, they also markedly increased the risk of getting hit by ground fire.

The assumptions of an uncontested attack proved tragically wrong. As the A-4s rolled in for their runs, the North Vietnamese were ready with 37mm and 57mm AAA fire. After completing his first run and beginning another pass flight leader George Wales spotted Jim in his parachute, descending close to the target. Jay Greene also spotted the top of Jim's parachute as he began lining up for his second pass. George flew by Graham at 800 feet, noting that he waved, signaling he was alive and seemingly unharmed. He came down in trees bordering the village of Kien Thon, which had mistakingly been bombed that same morning by F-4s. Attempts to contact him by radio were unsuccessful. He was on his 192nd combat mission when he was shot down. Of the loss, Jay Greene noted in his diary:

> When we realized someone had been hit and ejected we did a radio check, and in that way we determined it was Jim, the number four man and last in the attack order. During that interval of the radio check our flight became something of a scramble and I was well out over the water orbiting when I observed several grey puffs of 37mm and 57mm flak bursts near my aircraft. I immediately took evasive action and was stunned by the fact that these gunners could threaten me so far from the target. It was then I realized those guns were radar controlled.

After regrouping off the coast the three remaining Skyhawks completed a second run on the target, ensuring its destruction. They returned to *Enterprise* with heavy hearts, having lost one of their own.

However, after last seeing Jim alive, they had high hopes they would see him again after the conflict ended.

For the squadron, there was little time to absorb their first loss of the cruise. On 5 May VA-113 flew two recce missions to the area of the Son Ca River that runs south and west of Vinh. Using guns and bombs they attacked barges and other targets of opportunity, though none of any significance. On the second mission high winds caused the Skyhawk bombs to miss their marks except Bear Taylor, who scored several hits. While in the air VA-113 pilots overheard on the guard frequency of three USAF F-105s shot down while on missions to the railway yards at Yen Vien and army barracks at Ha Dong, both in the Hanoi area. All three Thud pilots were captured and became POWs until released during *Operation Homecoming* in 1973.

A massive Alpha strike to targets in Haiphong was canceled on 6 May due to bad weather. Further south an A-4 of VA-93 *Blue Blazers* from *Hancock* was shot down near Cap Falaise, 30 miles south of Thanh Hoa while on a recce mission. Pilot Lieutenant (junior grade) Robert Wideman call sign "Raven 310" was hit by AAA, causing the aircraft (BuNo. 151082) to roll uncontrollably, possibly from aileron damage. Wideman ejected and was captured and held in North Vietnamese captivity before being released during *Operation Homecoming*.

The breakneck pace of combat missions continued, with VA-113 pilots flying multiple missions in a single day, which depending on the rotation could start as late as noon or as early as midnight. The early morning of 7 May began with Jay Greene and Karl Lassey flying a recce mission to the Song Ma River south of Thanh Hoa. They bombed the Bi Thang Airfield, which was still under construction. They also bombed barges transporting supplies on the river. That afternoon Greene flew another recce to the same area with George Wales.

Thunderstorms canceled a midnight hop on 8 May. Later that morning VA-113 flew escort for an RA-3B Skywarrior reconnaissance aircraft, taking pictures from Cap Falaise to Cap Chao. Early the next morning of 9 May Jay Greene and Ted Bronson took off at 0400, working with an E-2A Hawkeye equipped with an IR sensor to locate trucks transporting military equipment on Route 1A. The weather turned marginal, resulting in the Skyhawks dropping their bomb loads at 20,000 feet through an overcast layer on the command of the E-2A, with indeterminable results. The weather did not improve much for another hop at 1000 that same morning, with multiple overcast cloud layers. XO Burnett, Karl Lassey, Mike McGraw, Jay Greene, and Pat Anderson found a bridge at Cape Mui Ron to attack. They handily destroyed the target with Mk.82 bombs.

A three-carrier Alpha strike was launched on 10 May against Haiphong TPP East with 26 aircraft from *Enterprise*. The plant had been attacked before, however, prior efforts failed to destroy the generators and boilers. Intelligence of defenses protecting the TPP included twenty-three 85mm AAA sites, thirteen 37/57mm AAA sites, and countless automatic weapons of varying calibers. There were numerous confirmed SAM sites and the target was well within range of several VPAF airfields which hosted both MiG-17 and MiG-21 aircraft.

Tom Scott, George Wales, Ted Bronson, Bud Johnson, Ernie Christensen, and Tom Holton of VA-113 along with six Skyhawks from sister squadron VA-56 took off from *Enterprise* with each aircraft carrying two WWII-era M-121 bombs, and one Mk.84 bomb. The pilots reported heavy AAA as well as five SAMs, which exploded underneath the strike force. There were no casualties or aircraft lost on the Haiphong strike, however, an attack on Kien An Airfield near Haiphong resulted in the loss of an A-4C of VA-94 *Mighty Shrikes* from *Hancock*.

The CAG of CVW-5, Commander Roger "Dutch" Netherland, was hit by an SA-2 as he was approaching the target. With his aircraft on fire and streaming jet fuel, he turned to get feet wet and return to the carrier. His Skyhawk (BuNo. 149509) made it 10 miles south of Haiphong before it rolled inverted and crashed into the sea. More than likely Netherland was badly wounded or killed by shrapnel from the missile. He had been CAG of CVW-5 for just six months.

Beautifully in formation.

On 18 May VA-113 suffered their second loss of the cruise. Lieutenant Robert "Nort" Naughton was leading a section on an armed reconnaissance mission north of Thanh Hoa. They were attacking a rail target near Doung Phong Thoung when Naughton was hit by AAA while in a dive which forced him to eject from his Skyhawk (BuNo. 147842). Though Lieutenant Jerry Greenamyer tried to protect Naughton with 20mm cannon fire, he was eventually captured and spent the duration of the conflict as a POW until released during *Operation Homecoming*. Naughton was on his 194th combat mission when he was shot down.

Also on 18 May the XO of VA-76, Commander Kenneth Cameron, was lost while attacking the Thuong Xa transshipment point 10 miles north of Vinh. While diving on the target Cameron, call sign "Sun Glass 683" was hit by AAA, forcing ejection from his A-4C (BuNo. 147816). Fellow POWs reported that Cameron was in poor mental and physical condition during his internment at the notorious Hanoi Hilton, enduring torture, beatings, and solitary confinement much of the time. Eventually, in a bid to stop the barbaric efforts of his captors to exploit and extract information from him, Cameron ate only enough food to barely keep himself alive. He reportedly passed away on 4 October 1970, with his remains repatriated on 6 March 1974. For heroism while a POW Commander Cameron was posthumously awarded the Navy Cross.

On Friday 19 May the Navy suffered the worst single-day loss of aircraft in the conflict, becoming known as Black Friday. A total of six aircraft were shot down, two each from carriers *Enterprise*, *Bon Homme Richard*, and *Kitty Hawk*. The day resulted in the loss of ten men, four of whom would never return home. VA-113 had participated in the Alpha strike on Van Dien where four of the six aircraft were lost, however, the squadron emerged intact.

Jay Greene had left on 12 May for Subic Bay to pick up an airplane, arriving back on the afternoon of 19 May. He had no idea of the happenings until tuning his radio while waiting to trap back aboard *Enterprise*. He wrote in his diary that evening:

> I took off from Subic Bay and flew back on the wing of an A-3 Skywarrior. When I got to the ship I was early for recovery and loitered at altitude for some time. I decided to switch to the attack channel to see what was going on.

I was stunned by what I heard. Total pandemonium! Two aircraft down and many SAMs in the air. The day turned out to be one of the blackest of the conflict for the U.S. Navy.

Undeterred by the previous day's losses, Saturday 20 May was another busy day on Yankee Station. Skipper Bennett led a SAM suppression mission in support of a re-strike on the Van Dien truck facility, along with Tom Brown, Tom Holton, and Bear Taylor. They flew ahead of the strike group to suppress four SAM sites thought to have shot down the two F-4s and one A-6 the previous day. Three of the sites were still active and were quickly redressed by VA-113, providing some degree of retribution. George Wales, Bob Brennock, Bud Johnson, and Jay Greene flew a recce mission to the San Ca River south of Vinh, dropping their bombs on barges. A subsequent scheduled Alpha strike by VA-113 on Kep Airfield was canceled due to bad weather.

Skipper Homer Smith of VA-212 on *Bon Homme Richard* once again led his A-4 squadron on a daring Walleye mission to bomb the heavily defended Bac Giang TPP, located 25 miles northeast of Hanoi. It was the second major TPP strike for VA-212 in as many days. Despite withering ground fire the *Rampant Raiders*[37] pulverized the facility, leaving twisted metal and fire in their wake. The numerous attacks on electrical infrastructure in the north ultimately resulted in Hanoi and Haiphong being plunged into darkness, reducing available electricity by 87%.

After delivering his weapons to the Bac Giang TPP Commander Smith was shot down by AAA. He was captured alive, however, according to another Navy POW he was tortured to death, dying on 21 May in Hanoi Hilton.

Smith was a highly respected officer and aviator who was on his second combat tour of Vietnam. A 1949 graduate of USNA, he was posthumously awarded the Navy Cross for the May 19-20 mission. He was also the recipient of a Silver Star, Legion of Merit, and four DFCs. Earlier in the year Smith had personally met with President Johnson in Guam to advocate the Walleye and its effectiveness in the precision destruction of targets. He had been scheduled to rotate out of command prior to 20 May, however, he delayed the changeover so he could lead this critical mission that ultimately cost him his life.

The relentless pace of Alpha strikes continued for CVW-9 and *Enterprise*. On 21 May strikes were once again flown to the Hanoi TPP and Van Dien truck facility as well as another strike on Kep Airfield by VA-113 aircraft armed with Bullpup missiles. Jay Greene, who flew on the mission with Bob Brennock, remarked in his diary that although three SAMs were fired at them, all missed by a wide margin. However, the black blotches from 85mm AAA were so thick that as a result he had limited visibility outside the cockpit. Throughout his two tours to Vietnam, he ranks the 21 May mission to Kep as the most dangerous of his time flying combat in the theatre.

Two VA-113 Skyhawks were assigned the task of flak suppression for the mission. As they flew over the airfield as many as ten SAMs were launched at the strike force. Tom Brown targeted a particularly threatening site at close range with a Bullpup, pressing his attack despite the constant and deadly ground fire. His missile permanently silenced it, resulting in a Silver Star.

In the following days, VA-113 flew recce missions, working with FACs to destroy targets of opportunity. On 24 May a ceasefire was in effect for Vesak, a festival commemorating Buddha. Although ceasefires had been broken numerous times in the past, the day was uneventful. No American aircraft were lost and VA-113 recce flights fired no ordnance.

After a rare quiet day over North Vietnam, the grueling pace picked up again on 24 May. Although inclement weather had ruled out any targets in Hanoi and Haiphong, Navy aircraft still had plenty of other targets in the southern route packages. Jay Greene and XO Andy Burnett departed for a night mission when Burnett experienced a landing gear malfunction, necessitating his return to the carrier. Greene then joined up with two A-4s of VA-56 and proceeded to an area near Long Son known as the "Shrimp," bombing a ferry.

Firing a Bullpup on a target.

Aircraft from the squadron launched again at 0800, with Jay Greene, Tom Scott, and Pat Anderson carrying Bullpup missiles. Their target was the Cam Pha railroad bridge east of Haiphong. The flight scored several hits resulting in damage, but as with many bridges attacked in North Vietnam, it was still standing when they departed the area.

Pilot Lieutenant (junior grade) M. Aslop was flying an A-4E call sign "Raven 311" (BuNo. 151076) of VA-93 from *Hancock* on a target 10 miles southwest of Ninh Binh when he was hit by AAA. Heading feet wet his engine began to fail, resulting in a flame-out. He ejected 15 miles from Thanh Hoa and was rescued by Navy helicopter.

Bear Taylor, Jay Greene, Pat Anderson, and George Wales flew a recce mission on 25 May to Route 15 west of Thanh Hoa, resulting in an attack against a bridge using Mk.82 bombs. While the bridge escaped largely unscathed the Skyhawks managed to crater the approaches, making it unusable in the short term. Southwest of Haiphong A-4s of VA-76 from *Bon Homme Richard* pounded the Kien An Airfield in a massive Alpha strike involving 27 Navy aircraft, 10 of which were A-4 attack aircraft while six A-4s were flying Iron Hand SAM suppression.

The missions to destroy the electrical grid of North Vietnam continued on 26 May, with another Alpha strike to Haiphong TPP West. Ten A-4s each carrying two Mk.82 bombs and one Mk.83 bomb hit the target, causing extensive damage. Kep Airfield was attacked again, resulting in the loss of an A-4E of VA-93 from *Hancock*. Lieutenant (junior grade) Read Mecleary call sign "Raven 300" (BuNo. 152022) had arrived over the target at an altitude of 13,000 feet and was assigned a flak suppression role for the strike. His aircraft was hit by AAA, causing a constant roll to the right. Turning to make feet wet back to the carrier, the struggle to keep the aircraft under control became untenable, forcing him to eject 12 miles east of the target. Mecleary, who was on his 56th combat mission, was captured. Taken to Hanoi Hilton, he was unable to walk for two months due to injuries suffered during the ejection. He was released during *Operation Homecoming*.

The last day of the line period for CVW-9 was 27 May. VA-113 flew several missions that day. The first was a recce mission to canals that ran north and west of Vinh. Tom Scott, Bill Bowes, Walt Lenhard, and Jay Greene dropped bombs on cargo barges on these ancillary waterways. Later that day Jay Greene and Bill Bowes were joined by Bear Taylor and Don Williams on a strike to the Dong Phong Thaung railroad bridge. The target was hit multiple times, however, the F-4 fighters provided flak suppression and TARCAP reported that after VA-113 left the area and the air cleared, the bridge was still standing.

*Enterprise* departed Yankee Station on 28 May, steaming for Subic Bay and well-deserved liberty for the crew. It had been a difficult month on the line, with the frequency and intensity of Alpha strikes steadily increasing. Though the crew still had one more line period before they returned stateside, light was beginning to appear at the end of the long tunnel of combat operations.

While *Enterprise* was in Subic Bay taking a break from combat operations, the pace continued unabated on Yankee Station. On 30 May Commander James Mehl was flying an A-4E call sign "Raven 301" (BuNo. 151049) of VA-93 from *Hancock* against the Do Xa transshipment point, 15 miles south of Hanoi. Mehl, who was the XO of VA-93, was flying an Iron Hand SAM suppression role. After barely evading one SAM Mehl climbed to 16,000 feet to fire a Shrike at a SAM radar. Another SA-2 suddenly appeared, streaking toward his aircraft. With little time to react the missile exploded near his Skyhawk. Mehl turned east to get feet wet but was unable to keep his badly damaged aircraft airborne. He ejected near Hung Yen and was captured, becoming a POW until his release during *Operation Homecoming*.

The last day of the month brought two additional losses for VA-212 from *Bon Homme Richard*, who were flying many Walleye missions to heavily defended Hanoi and Haiphong. On 31 May four A-4s from the squadron were heading to another raid on Kep Airfield when they encountered heavy AAA about 20 miles northeast of the target. Flight leader Lieutenant Commander Arvin Chauncey call sign "Flying Eagle 223" (BuNo. 151113) was hit by AAA in the engine, causing it to catch fire. He jettisoned his stores to try and get as much glide out of his Skyhawk as possible, however, the aircraft lost power and he was forced to eject feet dry 20 miles northeast of the target. He became a POW in Hanoi Hilton until released during *Operation Homecoming*.

When Chauncey went down the remaining aircraft in his flight began orbiting above to provide support for a rescue. Lieutenant (junior grade) M. Daniels call sign "Flying Eagle 229" (BuNo. 151183) took a hit from AAA eight miles northeast of the target which caused a massive fuel leak. He turned east to get feet wet and find a tanker, however, AAA had damaged his radio, preventing him from contacting other aircraft. Without enough fuel to reach the carrier, he had luckily found a destroyer when his engine flamed out. He ejected and was rescued by a Navy Seasprite helicopter. After taking heavy AAA fire and losing two aircraft before even reaching the target, the strike force returned to *Bonnie Dick*, giving Kep Airfield a temporary reprieve.

*Bon Homme Richard* and CVW-21 had suffered high losses during their WESTPAC, which had begun on 26 February. In just over three months, through 31 May, CVW-21 lost 13 aircraft resulting in 6 POWs and 6 fatalities. Two of those POWs, Commanders Robbins and Smith, died while in North Vietnamese custody.

On 3 June *Enterprise* departed Subic Bay at 1800 hours, en route back to Yankee Station. The following day VA-113 and other combat squadrons received briefings on events that occurred while they were off the line. Particular attention was given to the continuing attacks on Kep Airfield and other targets in the greater Hanoi and Haiphong areas.

*Enterprise* resumed combat operations on 5 June with 22 sorties, despite bad weather. Many within the air wing began paying close attention to the Six-Day War that had broken out in the Middle East between Israel and its neighbors. Though few believed the United States would become militarily involved, it nevertheless concerned many that an American Navy vessel, USS *Liberty* (AGTR-5), had been attacked by Israeli forces, resulting in 34 fatalities. Despite the attack on *Liberty* the U.S. and Israel remained close allies. A benefit of this alliance came shortly after the Six-Day War ended, when Israel turned over to American intelligence a fully intact SA-2 SAM system they had captured from Syria. The data gathered from this bounty of spoils enabled the U.S. to further develop effective countermeasures to the system.

Bear Taylor, Captain Richard DePrez (ops officer of CVW-9) and Tom Scott.

Post-BDA photo of Bullpup damage to railroad bridge in Vaih Hoc 6 June 1967.

An improvement in the weather on 6 June allowed VA-113 to fly a Bullpup mission to a railroad bridge in Vaih Hoc, southwest of Vinh. While the bridge remained standing after the attack, the tracks were badly damaged, putting it out of commission until the North Vietnamese could make repairs, which they often did at astonishing speed. *Enterprise* and other carriers on Yankee Station launched an Alpha strike later in the day against the Que Vinh vehicle depot ten miles south of Hanoi. However, the strike was canceled due to bad weather.

Yet another attack that day occurred on the SAM site near Van Dien, the location where the A-6 and F-4 from *Enterprise* were lost on 19 May, in the area known as SAM Alley. The site was finally destroyed, providing a slight degree of retribution. Of the successful attack Bear Taylor wrote in a letter to his wife:

We hit a missile storage area about 3 miles from where I had my duel with a SAM site on 19 May. A photo plane got a pix yesterday and a fine eye spotted ten missiles hidden in the bushes. We charged in under marginal weather and thank goodness no missiles came up at us, so we found the spot and hit it. Did we hit it! I put my 57 rockets right in there, along with almost everybody else, and we had missiles frizzling up through the air, skittering along the ground, and blowing up on their trucks. It was great. There were ten, we got nine and two radar vans, and an oil truck. There was some AAA but not enough to dim our glee. It was great fun.

On 8 June another mission was launched to SAM Alley with Jay Greene, Bear Taylor, Bob Brennock, and Pat Anderson, joined by five A-4s of VA-56, four A-6s of VA-35, and four F-4s. They were almost to the target when bad weather forced them back. They instead turned their weapons on barges. While egressing to feet wet over Thanh Hoa they encountered radar-controlled AAA which caused damage to one of the F-4s.

The following day, while A-4s of VA-212 from *Bon Homme Richard* were pounding a Hanoi TPP for a third time, Ted Bronson, Bill Bowes, Jay Greene, and Walt Lenhard flew an evening road recce mission west of Yen Thanh, an area known as the "Worm". There they found several trucks, but just as they were rolling in on the target AAA opened up, producing a vivid firestorm of tracers in the twilight sky. Jay Greene remarked in his diary:

We rolled in on it but an automatic weapons site opened up on us just as we were in our dive and it affected my bombing accuracy considerably. It was just starting to get dark and those tracers really show up. We got shot at several times during the hop.

An orange blossoming airburst was a tell-tale sign of an SA-2 SAM detonation.

Another massive Alpha strike was launched at the Van Dien missile storage and assembly area on the morning of 10 June. *Enterprise* launched eight A-4s of VA-113 armed with 2.75" rockets and Bullpup missiles while two A-4s of VA-56 were launched to provide Iron Hand SAM suppression. They were accompanied by F-4s and A-6s carrying Mk series bombs. Commander Peter Sherman, skipper of VA-56, and his wingman were 10 miles southwest of Hanoi when a volley of SAMs was fired at them. Taking evasive action, the wingman lost sight of

Sherman's A-4 call sign "Champion 406" (BuNo. 145145) and he was unable to reestablish communication. It is assumed 406 took a direct hit as no parachute was seen and no beacon or voice communication were received. Sherman had taken over as CO of VA-56 less than three months prior. His remains were repatriated and identified in 1991. Jay Greene was strapped into his aircraft on the flight deck that morning when Sherman walked past, on the way to man his A-4 for the strike. Later that night he noted in his diary:

> I had a Condition 5 RESCAP the day of this Alpha strike. Suited up and strapped in an A-4 cockpit more than ten feet above the carrier deck I watched as the pilots strode forth prior to manning their aircraft. Commander Sherman nearly walked under the nose of my plane. I later thought it ironic that he was smiling ear to ear as if he was on his way to a picnic or party.

Yet another Alpha strike was launched on 12 June, to the Dong Chan POL storage area 35 miles southwest of Haiphong. While VA-113 were hitting their target the pilots spotted a tall plume of black smoke in the distance, a result of *Bonnie Dick* aircraft attacking another nearby POL site. In a surprising but welcome development aircraft of VA-113 did not encounter any AAA or SAMs.

The day of 13 June brought several hops for VA-113. The first was a dawn mission, attacking roads south of Thanh Hoa with Mk.81 bombs. The second was a nineteen-aircraft strike against an ammunition storage facility in the port city of Hon Gai. The attacking contingent was composed of A-4s, A-6s, and F-4s. Jay Greene's ALQ-51 radar jammer was inoperative, leaving him to rely on Tom Brown to call out radar-tracked guns so he could take evasive maneuvers. The day marked one week left on the cruise, a milestone which was undoubtedly not lost on the air wing, who were feeling the fatigue from six months of flying combat. The end was near.

Getting ready to launch.

The draw-down of the cruise unfortunately did not equate to a draw-down of missions. At 0500 on 14 June VA-113 pilots were briefed for a recce mission to Route 18 north of Cam Pha. The Skyhawks were just 14 miles south of the border with China when they bombed a ferry and building. This area of the border was also being heavily worked daily by USAF fighter bombers from Thailand. Their targets were primarily railroad infrastructure as well as VPAF airfields.

The early morning hours of 16 June saw two VA-113 Skyhawks depart *Enterprise* at 0200, bombing under flares south of Thanh Hoa. Later that day VA-113 departed on another mission, a road recce between Cap Falaise and Thanh Hoa, ten miles west of the coast. Unable to find any trucks the pilots turned their attention to a bridge, with the bombs failing to find their mark.

Another early morning hop took place on 17 June, departing the carrier at 0230. No trucks were seen so bombs were dropped on a road, cratering it. A mission later that day to the Cam Pha causeway bridge was changed to a smaller railroad bridge south of Thai Hoa. Bud Johnson's Bullpup would not fire, while Tom Brown missed the wooden bridge by 80 feet and Jay Greene missed by 10 feet.

It was again before dawn on 18 June when a flight of *Stingers* took off on a road recce mission to Highway 1A. Jay Greene closed out his diary for the day remarking:

> Now that we have only two days left in this cruise you would think that we could relax a little—but tomorrow our squadron of 18 flyable pilots has 30 sorties! In addition, they are planning a large Alpha strike on Hai Dong. That's all we need—someone getting bagged in the last two days!

The Alpha strike of 18 June to Hai Dong was in an area of high activity between Haiphong and Hanoi. A total of 35 aircraft took part in the strike, with VA-113 targeting the Hai Dong Bridge while A-6s targeted the rail line to the north. Arriving feet dry north of Cam Pha, the strike force encountered heavy 37mm/57mm AAA fire. Several of the Skyhawks scored hits on the bridge while dodging two SAMs.

The stroke of midnight on 20 June saw the dawn of the last day of the cruise. Shortly after, at 0130, pilots of VA-113 took to the sky on a road recce mission of Route 1A, from Brandon Bay to Cigar Island. Looking to the north the early morning sky was lit up by 85mm AAA firing at Intruders attacking Dragon's Jaw Bridge in Thanh Hoa. With an absence of trucks on the highway, the pilots turned their attention to other targets, including the bridge that connected Cigar Island to the mainland.

Every aircraft that launched that day from *Enterprise* returned to the carrier unharmed, earning a tremendous sigh of relief from all. Getting "bagged" on the last day of a deployment is a fate dreaded by soldiers, sailors, and airmen since the earliest days of combat. Luckily, no man in VA-113 or CVW-9 would carry that distinction on this cruise. That evening more than one indiscreet alcoholic concoction was poured, leaving to many hangovers the next day when an awards ceremony was held while en route to Subic Bay. After two days in the Philippines, *Enterprise* departed for San Francisco, arriving at their home port of NAS Alameda on 6 July.

During a seven-month deployment, *Enterprise* spent 132 days on Yankee Station. The air wing launched 13,392 total sorties, dropping 14,023 tons of bombs on North Vietnam and Laos. The two A-4 squadrons on board flew 4,371 strike sorties, more than triple the number flown by F-4 and A-6 squadrons. Though the 14 aircraft lost during the cruise could be replaced, the loss of human lives was irreplaceable. In total ten naval aviators from *Enterprise* died in the line of duty, four were taken as POWs and one, VA-35 A-6 B/N Lieutenant Commander James Kelly Patterson, remains MIA. Of those losses, the *Stingers* accounted for one POW and one fatality. While any loss proved difficult, given that 1966-1967 represented the height of American air losses in SEA, VA-113 fared much better than many other Navy combat squadrons as a whole, a testament to their training, skill, tenacity, and teamwork. For their outstanding performance, VA-113 was awarded the Navy Unit Commendation medal.

Back at Lemoore the squadron once again underwent numerous personnel changes. Among those who were staying for the 1968 cruise were Pat Anderson, Ernie Christensen, Don Williams, Bear Taylor, Bob Brennock, Tom Brown, Michael McGraw, Walt Lenhard, Thomas Hilton, and Jay Greene. One new addition to the squadron for the 1968 cruise was nugget John "Jesse" James. Squadron mates remember James loved to fly, was fearless, and full of youthful exuberance. Bear Taylor remembers one mission late in the 1968 cruise where he and James destroyed a bridge southwest of Vinh using three Mk.83 bombs. Still armed with two AGM-12 Bullpup missiles each, Bear found three other bridges that could be used as a bypass to the one they just destroyed. Despite 37mm AAA opposition, they destroyed the bypass bridges. Of the mission, Bear recalls:

> Four bridges in one hour of great flying. Jesse had hits on three of them. I submitted his performance for DFC consideration, which was supported by Skipper Bob Thomas. Unfortunately, the Air Wing Awards Board turned down the recommendation.

On 3 January 1968, the *Stingers* of VA-113 once again departed NAS Alameda for Yankee Station. The year 1968 would prove to be a seminal one for the United States, marred by domestic political and social upheaval. There were also several international incidents, all of which truly tested the mettle of the country.

After spending a week in Pearl Harbor, *Enterprise* departed Hawaii for Sasebo, Japan. After three days in port, the boat left Sasebo at 0900 on 23 January with a destination of Subic Bay before arriving on Yankee Station. However, shortly after departing, the carrier received a message ordering them to hold at a navigational point some 100 miles west of Japan.

On 22 January North Korean vessels had begun harassing the U.S. Navy vessel *Pueblo*, which was conducting intelligence operations in international waters. The next day, some 15 nautical miles northeast of Wonsan, the *Pueblo* was confronted by armed North Korean vessels, which twice opened fire, resulting in the deaths of two U.S. Navy sailors. In a blow to America's prestige *Pueblo* ultimately surrendered and was boarded and seized by North Korea at 1455 Korean Standard Time (0555 UTC). For several weeks tensions ran high as another conflict on the Korean peninsula looked likely. *Enterprise*, some 600 miles to the south, was overflown by Soviet Tu-16 Badger bombers for several days.

Meanwhile, the air wing began prudently preparing for the possibility of a skirmish with North Korea. On 4 February the squadron began eyeing possible targets within the country and flew a practice Alpha strike the next day. Much to everybody's relief the crisis eventually de-escalated, and 338 days later the surviving 82 crew members crossed the DMZ into South Korea. The *Pueblo* was kept as a trophy and is now moored on the Taedong River in Pyongyang.

Exactly four weeks after CVW-9 departed aboard *Enterprise* for their third WESTPAC, on 31 January North Vietnam launched the Tet Offensive against the Republic of [South] Vietnam. The United States suffered yet another shock as South Vietnamese and American forces were caught off-guard in a massive offensive that reached deep into South Vietnam, with extensive commando and sapper operations by

NVA troops and Viet Cong guerrillas in major South Vietnamese cities including Da Nang, Saigon, and Hue. Ten days before the offensive the NVA launched several diversionary offensives along the DMZ, most notably at the Marine Corps forward fire base at Khe Sanh.

America and ally South Vietnam ultimately emerged victorious from the Tet Offensive, with the opposition suffering heavy losses. However, the shocking images of the battles broadcast on television stood in stark contradiction to how the Johnson Administration had been publicly portraying the conflict. The images of VC guerrillas infiltrating the U.S. Embassy in Saigon were especially troubling. The offensive turned a large bloc of the American public squarely against the effort. This was the beginning of the end of American military involvement in Indochina, as well as the sunset of the Johnson Administration. In a famous television address broadcast on 31 March, President Johnson chose not to run for re-election in 1968, irretrievably broken by the conflict that had consumed his administration, and the country.

On 16 February *Enterprise* was released from its holding point west of Sasebo and continued to Subic Bay. However, the following day the carrier and air wing were put on alert over the communist Chinese harassment of an American submarine. With the *Pueblo* incident still very fresh in their minds, *Enterprise* steamed to the aid of the submarine, with aircraft armed and pilots suited up. Wisely, the Chinese ended their provocation without any shots being fired.

Despite a tumultuous beginning, and the ferocity of fighting as part of the Tet Offensive, losses for VA-113 that 1968 cruise were one pilot and two aircraft. The first of these losses occurred on 7 May, when Lieutenant Commander Paul "Pete" Paine was flying RESCAP for an F-4B (BuNo. 151485) of VF-92 from *Enterprise*. Lieutenant Commander E.S. Christensen and RIO W.A. Kramer ejected north of Vinh after being shot down by a K-13 Atoll missile fired from a MiG-21, landing feet wet five miles off the coast. After the men were successfully rescued Paine headed back to *Enterprise*, only to be waived off his first landing attempt due to a fouled flight deck. While on downwind for a second attempt, his A-4 (BuNo. 154214) suddenly and inexplicably pitched nose-down and impacted the water. Though he ejected, it was too late, and outside the safety envelope. Paine, who was also an LSO for CVW-9, was killed instantly.

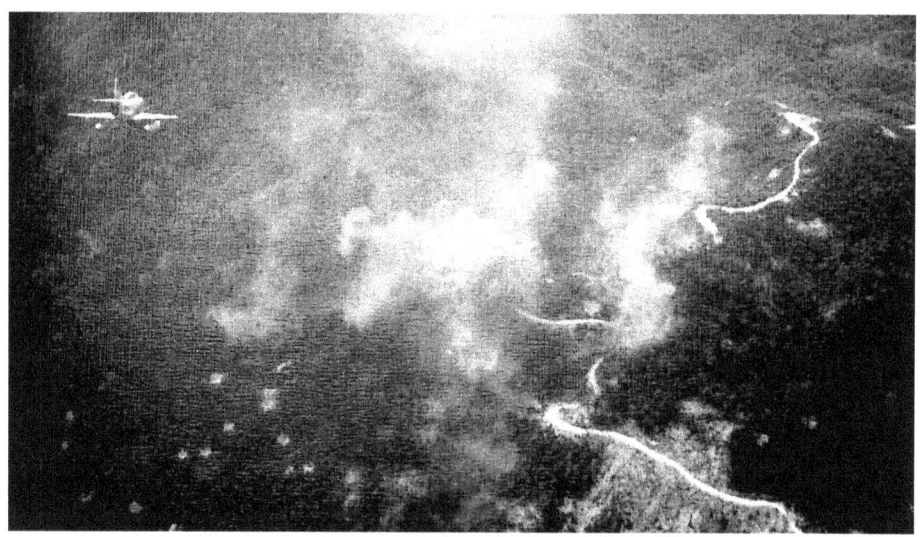

Pete Paine diving on a target 7 May 1968. This is the last known picture taken of him. He crashed into the Tonkin Gulf while preparing to recover aboard *Enterprise* shortly after this photo was taken.

The second loss occurred on 23 June when Ernie Christensen and Walt Lenhard flew a coordinated attack against a target in Vinh. After egressing from the area feet wet on the way back to *Enterprise* Christensen noticed his throttle control was not working. While troubleshooting the problem the engine in his Skyhawk (BuNo. 154216) quit, and subsequent attempts to re-light were unsuccessful.

He ejected about 20 miles from *Enterprise* and was quickly rescued by Navy helicopter. He later recalled that as he was descending by parachute into the Tonkin Gulf he suddenly developed a rabid fear of venomous sea snakes that are common to the waters that time of year. Once splashed down he climbed into his life raft in record time.

Jay Greene was also flying a sortie that day when he heard the mayday radio call from Walt Lenhard. He flew to the area where Christensen was floating in the sea, taking several candid snaps of the waterlogged aviator. He later remarked in his diary:

> Today I flew a coordinated strike on Vinh. On the way back Ernie Christensen's aircraft flamed out and he couldn't get a relight. He ejected 20 miles from the ship. I took pictures of him in the chute and his raft. He is okay. That's the second aircraft our squadron has lost on this cruise. On the next recovery, Pat Anderson collapsed a nose strut on landing. He was unhurt. Not too good a day for the *Battle Cry's*.

On 26 June VA-113 flew their last combat mission of the 1968 cruise. The *Stingers* were awarded a prestigious and coveted Battle Efficiency "E" for their outstanding achievements, as well as the Arleigh Burke Award. That summer, shortly after the conclusion of the 1968 cruise, John James was killed when his Corsair II impacted terrain while on a training flight from NAS Lemoore. Three years later, in November 1971, James "Zerbs" Zerblis, who had also joined the *Stingers* on the 1968 cruise as a nugget, died after ejecting from his Corsair II. He had been on a training flight with VA-205 *Green Falcons* from NAS Atlanta when his aircraft experienced an unrecoverable mechanical malfunction.

Although years would pass and American forces had long since withdrawn, there was still the matter of Jim Graham, who had been shot down on 4 May 1967 and last seen alive and seemingly unharmed after ejecting from his Skyhawk. When *Operation Homecoming* commenced in early 1973 Jim was not amongst the 591 men released, with North Vietnam claiming no knowledge of his fate. In 1978 he was declared KIA and in August 1985 partial remains were repatriated with no explanation from the Vietnamese government.

When Jim finally came home eighteen officers whom he had served with during his two tours with VA-113 attended a memorial service at the George Washington Chapel of Valley Forge National Park. For many of them, it was the first time they had seen each other since their combat cruises. However, even with the passage of many years, the deep connection they shared had not diminished. At the wake held at the Graham home following the memorial service, the muster included a swing around the room, with each *Stinger* telling a Jim Graham story. Though a somber occasion, Graham's family and shipmates used the opportunity to fondly remember Jim and rekindle the memories they had of him. Bear Taylor remarks "That was an amazing hour or so, especially for the Graham family and I suppose for many of the *Stingers*. For me, it was an event to remember—a sweet memory that reinforced the tie to my old shipmates".

The same year that Jim Graham was declared KIA the veterans of VA-113 suffered another loss. On 24 June 1978 Lieutenant Commander Pat Anderson was with VA-83 *Rampagers*, flying an A-7E Corsair during a Mediterranean cruise aboard *Forrestal*. Anderson, who had steadily risen through the ranks and was the operations officer of CVW-17, died when his A-7 crashed into the sea during dive-bombing exercises two miles off the south coast of Sicily. According to his wingman, during his second run on the target, Anderson began a steep dive and continued below the safe altitude for recovery. He attempted to pull up during the final moments of the dive, however, it was too late. His aircraft slammed into the water, tail first. An SH-3 Sea King from *Forrestal* retrieved his body, which was located beneath his life raft amidst a debris field created by the crash. Coincidently, the skipper of VA-83 at the time of the incident was Commander Bob Naughton, who had served with Anderson in VA-113 before being shot down and taken POW on 18 May 1967.

Pat's funeral was a notable affair in the nation's capital, given his family's close ties to the Navy, with burial at Arlington National Cemetery. Ernie Christensen, who considered Pat a brother, was the officer selected to present the flag that had draped the casket to his Father, retired Admiral George Anderson, Jr. He somberly recalls:

> I will never forget his eyes when I presented him with the flag. He looked right through me at that moment, and I could tell that he didn't understand why I was there, and Pat wasn't. Why wasn't I the one to fly into the waters, and not his son.

In the decades following their time in Vietnam, many *Stingers* would go on to great achievements within the Navy and beyond. Ten of the men ultimately made the rank of captain and three advanced to the rank of admiral. One of them even broke the barriers of space.

Ernie Christensen left Vietnam after the 1968 cruise with VA-113, serving a tour with the Navy's *Blue Angels* aerobatic team. He returned to Vietnam for a 1972-1973 cruise aboard *Enterprise* as part of F-4 Phantom squadron VF-142, including flying combat on the day the

Paris Peace Accords were signed. He would go on to several command positions including NFWS ("Top Gun"), the replenishment oiler USS *Kansas City* (AOR3), and the aircraft carrier *Ranger* during the 1991 Gulf War. He retired from the Navy in 1997 as a rear admiral, having flown 360 combat missions over three tours of Vietnam.

After serving two combat tours of Vietnam with VA-113 Bill Bowes returned for a 1971-1972 cruise aboard *Coral Sea* with A-7 Corsair II squadron VA-94. Bill went on to NTPS and Carrier Suitability Branch, working on important defense projects such as the Phoenix and Tomahawk missiles. He also served three years in the command ranks of VA-195 *Dambusters*. Bill retired from the Navy in 1996 as a vice admiral, having flown 350 combat missions over three tours of Vietnam.

Bear Taylor would go on to command an attack squadron, a carrier air wing, a replenishment oiler, the carrier *Coral Sea* and the Strike Fighting Wing of the Pacific Fleet. His final assignment was on the staff of CNO Admiral Frank Kelso II. He retired from the Navy in 1992 as a rear admiral, having accumulated more than 6,000 flight hours in 50 types and models of aircraft, over 1,000 carrier arrested landings, and more than 200 combat missions in the Vietnam theatre.

Don Williams returned to Vietnam for two additional combat cruises with VA-97, flying the A-7 Corsair II. He logged a total of 330 combat missions over four tours to Vietnam. In 1979 the former Navy ROTC cadet who was at first indifferent to aviation became a NASA astronaut, logging nearly twelve days in space as part of shuttle missions STS-51-D on *Discovery* and STS-34 on *Atlantis*. Williams became the second astronaut to emerge from VA-113 after Gene Cernan. Producing two astronauts was a feat wholly unique to the *Stingers*.

Wild Bill Ellis left the Navy in 1974 after twelve years and 186 combat missions over two tours of Vietnam. However, leaving the Navy was not necessarily Bill's choice. American forces were drawn down after the Vietnam conflict, leading to the departure of many combat veterans along with their experience and skills. After a decade of civilian life, including time as an airline pilot, Bill still felt the calling of the military. He began applying to re-enter service and it was a dogged task, however, his persistence paid off. In 1983 the Army accepted him at age 39 as an enlisted Noncommissioned Officer (NCO). He retired from the military in 1991 with twenty years of combined service between two branches.

Like several other aircraft of the Vietnam era, the Skyhawk continued to serve for many decades after. It served not only the United States but several key allies as well including Israel, Australia, and New Zealand. The Marine Corps retired the last of their Skyhawks in 1998, with the Navy following suit in 2003. Akin to many of its Vietnam-era contemporaries, the aircraft faithfully performed its duty for nearly half a century. It leaves a proud and lasting legacy and continues to be recognized and celebrated as the most effective and influential Navy light attack aircraft to ever grace the skies.

---

[25] The Navy lost 272 A-4 Skyhawks in the Vietnam conflict, while losing 128 F-4 Phantoms. Navy A-4 losses by year: 1964-7 1965-43 1966-77 1967-77 1968-32 1969-16 1970-12 1971-1 1972-7. These numbers represent both combat and operational losses.

[26] The first Navy loss of aircraft and the lives aboard in the Vietnam theatre occurred on 17 February 1960. A Douglas R4D-5/C-47 Skytrain (BuNo. 17244) was on a flight from Saigon to Hue when it flew into Hon Chay Mountain near Da Nang. Lieutenant Commander George W. Alexander, Lieutenant Commander Roger H. "Moon" Mullins and Avionics Electronics Technician Chief (ATC) William M. Newton were killed. While the exact cause of the incident remains unknown, enemy activity is not suspected.

---

[27] A total of 18 different Navy aircraft carriers would maintain a constant presence on Yankee Station off the east coast of North Vietnam from April 1964 to December 1973. There were normally three aircraft carriers on station, operating on a rotational schedule 24 hours a day.

[28] CIA-owned *Air America* maintained a large presence in Indochina with a fleet of aircraft comparable to a regional airline, flying classified missions throughout both Vietnams as well as Cambodia and Laos. The CIA also owned *Civil Air Transport*, which clandestinely transported supplies and troops to the French during the siege at Dien Bien Phu. Two CIA pilots, James McGovern, Jr. and Wallace Buford, were killed when their Fairchild C-119 "Flying Boxcar" was shot down 6 May 1954, the day before French forces surrendered to the Viet Minh.

[29] The P-4 was manufactured and supplied by the People's Republic of China (PRC), who delivered a dozen to North Vietnam between 1961 and 1964. Fast, maneuverable and featuring aluminum hulls, in the Chinese version Soviet-supplied 12.7 mm machine guns were replaced with higher-caliber 14.5 mm guns. The Chinese P-4 was based upon the Soviet-designed *Komsomolets-123*.

[30] The attacks on *Maddox* and *Turner Joy* in the early morning hours of 4 August are alleged and in dispute. Regardless, allegations of a second attack was used by the Johnson Administration as further justification for military escalation in Indochina via the Gulf of Tonkin Resolution.

[31] The address to the nation by President Johnson preceded the attacks by two hours, giving the North Vietnamese ample notice to prepare. The poor timing was due to a misunderstanding between CINCPC Admiral Ulysses Grant Sharp and Secretary of Defense Robert McNamara.

[32] Between 23 March 1961 and *Pierce Arrow* 5 August 1964 the U.S. military lost 45 aircraft in theatre, resulting in 78 fatalities. Of these numbers the Navy accounted for 4 aircraft lost (2 RF-8, 1 A-4, 1 A-1).

[33] Sears, David. *Such Men as These*. Da Capo Press 2010.

[34] On 18 April 1942 Captain Marc Mitscher was CO of *Hornet* when 16 B-25J Mitchell bombers of the Army Air Forces, forever known as the "Doolittle Raiders," took off from its flight deck and bombed Tokyo just four months after Pearl Harbor. It was this action that prompted Japan to attempt the invasion of Midway, a move which proved disastrous for the Imperial Japanese Navy, and one from which they would never fully recover.

[35] Stephen Coonts and Barrett Tillman detail this attack in their book *Dragon's Jaw: An Epic Story of Courage and Tenacity in Vietnam* Da Capo Press 2019.

[36] The Corsair II was also known as a "SLUF," an acronym for "Short Little Ugly F**k".

[37] An excellent book on VA-212 and their 1966-1967 deployment onboard *Bon Homme Richard* is *Rampant Raider: An A-4 Skyhawk Pilot in Vietnam* (Naval Institute Press) Stephen Gray 2007.

# Chapter IV

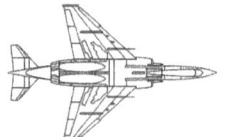

# It's Showtime
## The Continuing Saga of Ensign Thornton

It was a bitter disappointment for McDonnell Aircraft when, in May 1953, the Navy chose the Chance-Vought F-8U Crusader as their new high-speed air superiority interceptor over McDonnell's proposed upgrade of the company's F-3H Demon. There was good reason for dismay among the McDonnell team. During the evaluation phase, the Navy had changed requirements several times, including mandating an engine change that put the McDonnell design at a pronounced disadvantage. Despite losing out on the lucrative contract, they firmly believed their design to be superior and set out to prove it. They canvassed, questioned, interviewed, and surveyed every member of the Navy that was willing to talk. Out of these efforts came a trove of information, of not only what the Navy currently needed, but of developing future requirements as well; a practical wish list. Seemingly peering into the future the McDonnell team came up with a design they felt met all current and future needs of the Navy.

On 19 September 1953, just four months after the upgraded Demon was rejected in favor of the Crusader, McDonnell presented an unsolicited proposal for the F3H-G Super Demon. It outlined an all-weather, single-pilot, twin-engine interceptor and attack fighter. At that time, having just finished an exhaustive process that resulted in the selection of the F-8, the Navy was content with the Crusader in the interceptor role and the Douglas A-4 Skyhawk filling the attack role. However, McDonnell's research and tenacity impressed the Navy, especially in the promised top speed approaching Mach 2. That made it faster than any other Navy aircraft of the time, including the Crusader. A full-scale model of the F3H-G was ordered for further study of the proposal. On 18 October 1954, McDonnell received a letter of intent from the Navy for two prototypes. They were configured as twin-engine, 20mm cannon-armed, all-weather single-pilot fighter designated YAH-1.

Seven months later, on 26 May 1955, four Navy officers met with McDonnell Aircraft and presented a bid proposal for a new fighter. While the attack, all-weather and twin-engine capabilities remained, the Navy added the requirement that the YAH-1, above all else, needed to fulfill the role of a high-speed, missile-armed fleet interceptor. Therefore, instead of traditional internal cannons historically found on fighters such as the Crusader, the requirements specified four AAM-N-6 Sparrow air-to-air missiles and a powerful air intercept radar. The radar would be operated by a dedicated officer, necessitating the addition of a back seat. The engines were also changed, from Wright J-65 to General Electric J-79-GE-8. Also specified was a patrol range of 250 nautical miles from a carrier, and staying airborne for as long as three hours without the need for refueling. McDonnell was able to meet these requirements, and the YAH-1 became the XF4H-1.

As testing and evaluation progressed on the prototype numerous changes were made to the design. Wind tunnel testing resulted in a 12° increase in dihedral (up) to the outer portion of the wings to address lateral instability and a 23° decrease in anhedral (down) to the tailplane as well as a "dogtooth" wing feature to improve control at high angles of attack, giving the Phantom II its distinct appearance. To maximize airspeed, intakes for the powerful engines were reworked, with one fixed ramp and one variable geometry ramp to deliver maximum performance during supersonic flight. The nose and main landing gears were also enhanced to better withstand the unique challenges and rigors of operation from an aircraft carrier. While early suggested names for the new aircraft included Satan and Mithras, the less-controversial name Phantom II, later shortened to simply Phantom, was ultimately chosen, in homage to McDonnell's first jet-powered Navy fighter aircraft, the Phantom I. It had entered service in August 1947 as the first purely jet-powered aircraft to operate from an aircraft carrier. The Phantom I had a very short life, retired in July 1954 and replaced by the McDonnell F2H Banshee and Grumman F9F Panther.

On 25 July 1955, the Navy ordered two XF4H-1 test aircraft and five YF4H-1 pre-production examples. The first operational prototype, flown by test pilot Robert Little, took to the skies on 27 May 1958 from Lambert Field in St. Louis. This early testing resulted in further improvements, chiefly optimization for supersonic flight. That included the addition of 12,500 micro holes to each air intake ramp to bleed off the slower-moving surface boundary layer. Splitter plates were also added to divide and divert the boundary layer away from the intake ramps.

Chance-Vought had also been working on a new design to fulfill the requirements outlined to McDonnell by the Navy in May 1955 and had submitted a test prototype of the XF8U-3 Crusader III, an evolution of the F-8 Crusader II. By mandate of United States Congress the two aircraft were pitted against each other in open competition to determine who would win the contract to build the next-generation Navy fighter. The Crusader III performed impressively, hitting Mach 2.39 with an expectation that, after further testing and refinement, the aircraft could near a top speed approaching Mach 3. However, speed aside, the Navy was concerned with both the single-engine and lack of dedicated radar operator. On 17 December 1958, the Phantom was declared winner. The aircraft began carrier suitability testing, with the first launch and recovery cycle completed on 16 February 1960 from *Independence*.

The Navy quickly realized they had a remarkable aircraft in the F-4. Within two years of its introduction, the Phantom had broken no less than fifteen aviation world records for speed and altitude. By the time the first F-4As arrived to RAG training squadron VF-121 *Peacemakers* at NAS Miramar, the Phantom was already becoming a legend.

The F-4 was well-received within the Navy, with pilots and RIOs intrinsically drawn to the battle tank that happened to have wings and jet engines. It was tough, versatile, and most important of all, it was powerful and fast. The aircraft was so impressive that the Air Force, fiercely protective of their fighter doctrines, began jealously peering over the fence at the new toy the Navy was parking in their driveway. The Air Force quickly adopted the F-4 in 1963 in a fighter-bomber role, becoming the largest single user of the aircraft, making it ubiquitous over the skies of SEA. The first USAF Phantoms arrived in Vietnam in December 1964, delivered to the 43rd TFS in Cam Ranh Bay, Republic of [South] Vietnam.

Unavoidable within the realms of military aviation, the Phantom quickly acquired a slew of friendly nicknames, mostly centered on its large size and distinct appearance. The aircraft was commonly referred to as "Double Ugly," "Rhino," "Snoopy," and "Flying Anvil". Perhaps the most common was "Old Smokey" due to the thick, black columns of jet exhaust that followed wherever it flew.

On 8 July 1961 at NAS Oceana VF-74 *Be-Devilers* became the first squadron to receive the refined F-4B with the upgraded Westinghouse APQ-72 radar. They completed carrier qualifications in October and embarked with the Phantom on an August 1962-March 1963 Mediterranean cruise aboard *Forrestal,* the aircraft's first full carrier deployment. On 12 January 1962, VF-102 *Diamondbacks* flew the F-4B aboard *Enterprise* for the nuclear carrier's shakedown cruise, putting both ship and aircraft through rigorous paces. In September 1962 VF-114 took their F-4Bs aboard *Kitty Hawk* for a WESTPAC cruise.

It was on 5 August 1964 that Navy Phantoms made their first combat appearance over the skies of North Vietnam during *Operation Pierce Arrow*, which were limited retaliatory strikes in response to a patrol boat attack against the destroyer *Maddox*. This was followed two days later by an alleged second attack on *Maddox* as well as *Turner Joy*, now known as the Gulf of Tonkin Incident.

During *Pierce Arrow* F-4B Phantoms of VF-142 and VF-143 from *Constellation* provided combat air patrol for Douglas A-1 Skyraiders and A-4 Skyhawks making the attacks. Like a new employee on their first day on the job, the Phantoms passively observed, perhaps taking note of what was to come. Though at that time it was in the shadows of the burgeoning conflict, the F-4 Phantom would soon take centerstage for all the world to see.

With the presidential election only three months away President Johnson had tried to tread lightly in *Pierce Arrow*, however, the Gulf of Tonkin Incident was nevertheless the spark that ignited the powder keg of American military escalation in Indochina. The Gulf of Tonkin Resolution passed by Congress on 10 August 1964 was followed by the commencement of *Operation Rolling Thunder* the following March. Thus the United States plunged head-first into a regional conflict that would, in a short time, consume, divide and nearly destroy the fabric of the country.

On the same day *Operation Pierce Arrow* was being carried out in North Vietnam, Navy Fighter Squadron VF-96 call sign "Showtime" departed NAS Alameda aboard *Ranger* for their first combat deployment to Vietnam. During transit, the carriers *Bon Homme Richard*, *Ticonderoga*, and *Constellation* were launching multiple daily sorties in targets above and below the DMZ that separated the two countries.

In the first two years of the conflict, 1964-1966, carriers began their presence in Vietnam with a short stay of one week at Dixie Station, off the southeast coast of South Vietnam. Beginning combat sorties in the south allowed airmen time to hone their skills and acclimate to the tropical environment before flying from Yankee Station in the north, where the air defenses were much more modern and numerous.

*Ranger* and CVW-9 arrived to Dixie Station as part of CTF-77 in November, flying missions into South Vietnam. That same month the Soviet Union strengthened ties with the DRV and with the VC operating in the south. Among the military assistance granted to North Vietnam was the S-75 (SA-2) *Dvina* air defense system as well as 36 MiG-15 and MiG-17 fighter aircraft for the fledgling VPAF. The Soviets also increased aid in terms of arms and air defense systems, while China sent combat engineers.

Within the same time frame, the political situation in South Vietnam had rapidly deteriorated. Shortly after *Pierce Arrow* the country was thrown into crisis with an attempted overthrow of the government. The coup that followed in December, the first of four such political crises that would occur in South Vietnam within a year, further destabilized the situation. It also reinforced the feeling within many Americans that the government of South Vietnam was not a stable nor trustworthy partner.

The first stages of *Operation Rolling Thunder* commenced on 2 March, with a focus on ammunition depots, radar sites, and other military targets. On this first day, the Air Force flew missions against the ammunition depot at Xom Biang and the Quang Khe Naval Base. A massive armada of 130 aircraft took part in the strike. Though the target was destroyed, in a foretelling of the future heavy air defenses cost the Air Force five aircraft. Four of the pilots were rescued, while one was taken as a POW. Originally scheduled to last eight weeks, *Operation Rolling Thunder* instead lasted forty-five months, just short of four years. Ironically this was the same amount of time it took the United States to mobilize, fight and achieve victory in WWII.

As spring arrived weather began to improve, allowing the full force of *Rolling Thunder* to be unleashed upon North Vietnam. In addition to Navy carriers on Yankee Station the Air Force was flying daily sorties into the country from bases in Thailand, necessitating them traversing the no-mans land of Laos to reach their targets and return to base.

The first Navy Alpha strike of *Rolling Thunder* occurred on 15 March 1965, with aircraft from *Ranger* and *Hancock* attacking the ammunition depot at Phu Qui, located between Vinh and Thanh Hoa. A total of 94 aircraft took part in the strike, with the main force consisting of 64 A-4 Skyhawks and A-1 Skyraiders, with eight F-8 Crusaders providing flak suppression. Two RF-8s provided post-BDA intelligence, and eight F-4s, ten F-8s, and two A-1s provided CAP. Post-BDA photographs from the RF-8s showed 21 buildings severely damaged or destroyed in the strike. One Skyraider[38] was lost on the mission, resulting in the death of the pilot.

The first week of April saw the Air Force and Navy jointly launch large-scale coordinated strikes into North Vietnam. The targets were several bridges in the Thanh Hoa area, most notably the strategically crucial Dragon's Jaw Bridge over the Song Ma River. The first attack on the bridge occurred on 3 April, with the Air Force sending 46 F-105 Thunderchiefs, also known as "Thuds," and 21 F-100 Super Sabres, supported by 10 KC-135 Stratotankers. Two RF-101C Voodoo reconnaissance aircraft were used for pre and post-BDA. The Navy did their part with 35 A-4 Skyhawks, 16 F-8 Crusaders, and four F-4s. During the sortie three VPAF MiG-17s attacked, damaging a Navy F-8 of VF-211 from *Hancock* piloted by Lieutenant Commander Spence Thomas, causing a diversion to Da Nang. Despite the massive scale of the strike and the loss of two aircraft,[39] post-BDA photographs showed very little damage inflicted on Dragon's Jaw. The Air Force launched a second wave of attacks the following day, consisting of 58 aircraft. This second attack resulted in the loss of four additional aircraft,[40] with one POW and three fatalities. Two of these fatalities, both F-105 Thud pilots, were lost to VPAF MiGs. Despite sustaining more than 300 direct hits from bombs and Bullpup missiles over two days of attacks, there was only minimal damage, which was quickly repaired and the bridge remained in service. All that was to show for the loss of six aircraft and four lives were pockmarks and shallow craters on the piers and abutments.

The missions of 3-4 April are significant for several reasons. It was the first large-scale, coordinated attack on targets well within North Vietnam. It was also the first of many attacks against Dragon's Jaw Bridge. Lastly, it was the first direct engagement with VPAF MiGs.

The following day, 5 April, a Navy RF-8 reconnaissance aircraft took the first pictures of an SA-2 SAM installation under construction in North Vietnam, 15 miles southeast of Hanoi. Permission was requested to attack the site, however, the Johnson Administration refused, fearful that Soviet military advisors could be among the potential casualties.

Despite a non-stop flow of military equipment provided by the Soviet Union and China, including the SA-2 SAM, attacks on Haiphong, particularly the harbor, were strictly prohibited. The prevailing fear within the Johnson Administration was that any attack on the harbor could damage Soviet or other foreign vessels, or harm their personnel. Many American airmen who flew over Haiphong witnessed military hardware being unloaded from these ships, frustratingly powerless to intervene. A Navy F-4 RIO of the time notes:

> We would follow the [Red] river up to Haiphong, flying so low we would leave rooster tails in the water as we passed over. We would buzz the Soviet ships in the harbor, low enough to see Russian sailors sunbathing on the decks. They would leisurely watch us fly by, knowing we could do nothing.

Incredulously, a point came when the Soviet Ambassador to the United States lodged a diplomatic protest over American aircraft buzzing the ships, resulting in low-level flights over the harbor being prohibited.

Communist ally China also provided military training and equipment to North Vietnam, some shipped via rail across the border shared by the two countries. However, that rail line was also off-limits, as Chinese citizenry could be onboard. Despite Johnson's concern of provoking China, the Chinese would display a pattern of provocation throughout the conflict.[41] This was the case on 9 April 1965, resulting in the first combat loss of an F-4,[42] the first loss of the conflict for VF-96, and the first confirmed Navy[43] MiG kill of the conflict.

Carriers operating on Yankee Station routinely had a pair of Phantoms flying CAP in the northern portion of the Tonkin Gulf to protect the fleet against any sea or airborne threats such as patrol boats or enemy aircraft. With the end of their patrol approaching two VF-96 Phantoms were readied to launch from *Ranger* to relieve them.

As the first relief Phantom (BuNo. 151425) catapulted from *Ranger* its right engine failed, sending it plunging into the sea. Pilot Lieutenant Commander William Greer and RIO Lieutenant (junior grade) Richard "Bruno" Bruning ejected just as the aircraft impacted the water, and were rescued.

The second relief Phantom (BuNo. 151403) piloted by Lieutenant (junior grade) Terence Murphy and RIO Ensign Ronald Fegan call sign "Showtime 602" immediately launched, joined by an additional F-4 piloted by Lieutenant Howard Watkins and RIO Lieutenant (junior grade) John "Jack" Mueller. As the pair of aircraft flew north to relieve the two Phantoms on station they were ambushed by Chinese MiG-17 fighters. Murphy and Fegan made a radio call alerting them to the presence of the MiGs, resulting in the Phantoms that were standing by to be relieved immediately flying south to assist.

The ensuing fight took place at a high altitude south of China's Hainan Island. During the short but intense battle, Showtime 602 shot down[44] a MiG-17 using an AIM-7 Sparrow missile. Following the engagement the Phantoms regrouped to tanker, however, Murphy and Fegan were nowhere to be found. The last contact from them was a radio call announcing that they were out of missiles, and returning to the carrier. The Navy theorizes that the men were shot down by one of the MiGs, however, a Chinese newspaper claimed they were a victim of friendly fire, shot down by a Sparrow fired from a fellow Phantom. What ultimately happened to the men remains a mystery.

The 1964-1965 deployment was cut short after 103 days on the line, due to a major shipboard fire on 13 April, the result of a broken fuel line. The inferno consumed one of the carrier's main machinery rooms, resulting in one fatality. *Ranger* returned to Subic Bay on 15 April, and five days later steamed to NAS Alameda, arriving on 6 May. The squadron returned to Miramar while *Ranger* spent nearly five months in San Francisco Naval Shipyard undergoing repairs. The boat left the shipyard on 30 September and departed for another WESTPAC in December of that year with CVW-14.

During that shortened Vietnam cruise *Ranger* and CVW-9 lost five aircraft. In addition to the two Phantoms from VF-96 two A-1 Skyraiders were lost from VA-95 *Green Lizards* resulting in two fatalities.[45] An RA-5C Vigilante was also lost from RVAH-5, resulting in the deaths of pilot Lieutenant Commander Donald Beard and RAN Lieutenant (junior grade) Brian Cronin. This was the first loss of the conflict for the *Savage Sons* of RVAH-5.

A little over two months after *Ranger* and CVW-9 departed Vietnam, another significant event occurred. On 24 July Air Force pilot Captain Roscoe Fobair and WSO Captain Richard Keirn[46] of the 47th TFS/ 15th TFW from Ubon was one of four F-4C Phantoms on a CAP mission. They were providing cover to a flight of F-105 Thuds striking a munitions factory at Lang Chi, some sixty miles southwest of Hanoi. Seconds after spotting a smoke trail heading from the ground to their aircraft, an SA-2 SAM hit the Phantom (Serial No. 63-7599), forcing their ejection. This resulted in the two men having the unfortunate distinction of being the first American aircraft shot down by a SAM over North Vietnam. Keirn was released during *Operation Homecoming*. Despite a belief that Fobair also safely ejected, he was not among those released after the conflict. His partial remains, along with some personal effects, were recovered in 2001.

This first attack by VPAF MiGs on 3 April, along with the first loss of an aircraft to a SAM on 24 July contributed to a total of 375 aircraft lost in the Vietnam theater in 1965, resulting in 53 POWs and 347 fatalities. This produced a stark realization that the air war in Vietnam would be much more challenging, and deadly than previously thought.

In August 1965, three months after returning from Vietnam, VF-96 flew to Norfolk, departing on *Enterprise* for a month-long shakedown cruise to Guantanamo Bay and the Caribbean, with Commander Bob Norman as skipper. The men in the squadron were very fond of Norman, and describe him as an easy-going and personable man with whom everybody got along with.

After the shakedown cruise, notification was received that in less than two months VF-96 would be deploying again to Vietnam, aboard *Enterprise*. This came as a shock to many in the squadron, who were not expecting another WESTPAC deployment until the following year. Word was that Secretary of Defense Robert McNamara wanted to fulfill his pledge of having a nuclear-powered aircraft carrier on the line by Christmas. At that time *Enterprise* was the only carrier in the fleet that fit the bill.

As was routine with combat squadrons of the time, VF-96 experienced a large turnover in pilots and RIOs between deployments. Most of the new faces that summer came from RAG squadron VF-121 in Miramar. Being that *Pierce Arrow* missions were the first large-scale combat sorties for the U.S. Navy since Korea, very few of these new arrivals were experienced, combat airmen. Thirteen of them were nuggets, dependent upon veterans to learn the skills and knowledge necessary to survive. The most expedient method was to pair experienced pilots with inexperienced RIOs, and experienced RIOs with inexperienced pilots, when possible.

Navy Phantom pilots and RIOs shared a special relationship. Unlike the venerable Grumman A-6 Intruder, where pilot and B/N sat side-by-side, the crew of a Phantom was a front-and-back configuration. The RIO in the backseat had very limited forward visibility and no flight controls. Though the pilot and RIO could verbally communicate over the intercom system, the two could only physically see each other's faces via a mirror on the left-hand side of the pilot's cockpit. Yet, they had to work together seamlessly, sight unseen, in absolute tandem. Both pilot and RIO had to have a high amount of trust, faith, and confidence in each other. Their lives depended on it.

The term Radar Intercept Officer was a misnomer for Navy Phantom RIOs in the Vietnam theatre. They were much more. The RIO was a de facto co-pilot minus flight controls, and just as essential to the mission as the pilot. As the internal navigation system in the F-4B was notoriously unreliable, the RIO bore responsibility for this critical task. In addition to dead reckoning and paper charts, the RIO used land patterns on the radar to guide the aircraft safely to the target and back. In fact, he used his radar more as a navigational tool than one of combat. In addition, the RIO also handled radio communications.

Pilot Lieutenant (junior grade) James "Jim" Ritchie arrived to VF-96 in the summer of 1965, where he would serve his first tour flying with RIO Lieutenant (junior grade) Robert "Roach" Kania. Ritchie had entered the Navy from the ROTC pipeline before qualifying for flight training and earning his wings in January 1965. He recalls his youthful exuberance:

> Like many naval aviators on their first combat deployment to Vietnam, I was bright-eyed and ready to conquer the world. Going overseas and fighting communists in an exotic, foreign land was an adventure, and it was exciting! It is precisely why I joined the Navy.

RIO Lieutenant (junior grade) Doug Kindseth had arrived to VF-96 as a nugget in late March 1965, shortly before the fire that cut that deployment aboard *Ranger* short. He met the carrier in Subic Bay during a break from the line period before the cruise prematurely ended. He remembers of that time:

> I wasn't long out of flight training in VF-121 when I received orders to VF-96 and Vietnam, which at the time was fine with me. I was young, full of piss and vinegar, and wanted to be the tip of the spear. The reality of a few combat missions crushed that rather naive bravado.

Lieutenant (junior grade) Dave Hoffman arrived to the squadron that summer after a tour as a T-28 Trojan instructor. A 1962 graduate of USNA, he was a highly talented pilot who would one day become CAG of CVW-8 on USS *Carl Vinson* (CVN-70). He would also one day command USS *Nimitz* (CVN-68) as well as *Kitty Hawk*. Despite flying dozens of aircraft types over his long Navy career, it is the F-4 that holds a special place in his heart. Dave succinctly states "The Phantom was a great airplane and like a first girlfriend, you always remember it and always love it".

Dave flew with Lieutenant (junior grade) Vito "Beans" Pinto for a portion of the 1965-1966 cruise before he began flying with Doug Kindseth, who is hesitant to use the term "enjoy" to describe flying with Hoffman, or any pilot he flew with in combat:

> It is hard to use the term "enjoy" when you are getting shot at every day. However, I will say I was confident flying with Dave as he was an excellent pilot. I always knew with him in the front seat, as long as we weren't hit somewhere critical, we were going to survive.

RIO Ensign Paul Daley also arrived to VF-96 in the summer of 1965, just before embarking on his first deployment with *Enterprise* on the Caribbean shakedown cruise. He flew with Lieutenant Commander Norton "Windy" Winchester on his first combat cruise and recalls a reassuring talk Skipper Norman gave to the squadron before their impending WESTPAC:

I remember Skipper Norman standing in front of the squadron and stating that not one target in North Vietnam was worth anybody's life. He decreed caution and urged always choosing the safest course of action. He also said his primary responsibility as skipper was to make sure that everybody who left with the squadron came home with the squadron.

Pilot Lieutenant (junior grade) Richard "Rich" Wilson arrived to VF-96 in May 1965, just after the squadron's return from the cruise on *Ranger*. A 1963 graduate of USNA, he was on his first of what would be four combat deployments to Vietnam. Wilson would precipitously rise in the ranks, eventually commanding the carrier USS *Midway* (CVA-41), and as rear admiral commanded the *Nimitz* carrier battle group. His RIO on that first cruise was Lieutenant Jack Mueller.

Rich Wilson and Jim Ritchie (foreground) in ready room.

Wilson was a humble man with a keen sense of humor who quickly made close friends within the squadron. He was dependable, true to his word, and always there for his shipmates. Jim Ritchie roomed with Wilson on both the 1965-1966 and 1966-1967 cruises and remembers him as an outstanding aviator and valued squadron mate:

> Rich and I were quite different I suppose, in that I (almost) never thought of myself as a career naval officer, and he always did. He was married; I was not. Rich was a graduate of the Naval Academy, while I came through ROTC. Despite our different backgrounds we got to know each other very well and became close friends. He was a reliable, sound, and skilled pilot, never shying away from the most challenging of missions. I believe it was this, along with his superlative flying skills, that led the skipper to wisely choose him as his wingman on the subsequent cruise.

RIO Ensign Jim Stillinger arrived to Miramar and VF-96 with Rich Wilson in May. He flew with pilot Lieutenant Commander Arthur "Art" Cisson for the first months of the deployment. When Cisson left the squadron he began flying with other pilots. His most vivid memory of flying in an F-4 came before his first combat deployment when he was with RAG VF-121 in Miramar. On a high-altitude intercept training exercise, pilot and RIO decided to push the envelope of the aircraft to the edge of space:

> After the exercise, we had some fuel left. Since we were wearing pressure suits we decided to see just how high the

Phantom could fly. The pilot went full afterburner and pointed the nose of the aircraft straight up. As we rocketed past 60,000 feet he shut down the engines, letting momentum take us to an altitude of nearly 80,000 feet above the earth. As the aircraft reached the apex of the climb the curvature of the planet became visible, and the sky above disappeared, replaced by the blackness of space. It was a beautiful, wondrous site. As we began to descend and gain speed the engines quickly windmilled back to life and we returned safely to Miramar.

The *Fighting Falcons* departed Norfolk aboard *Enterprise* on 26 October 1965 for their second WESTPAC deployment to the Vietnam theatre. After ORI in the Caribbean, the carrier steamed east, around the Cape of Good Hope, into the Indian Ocean. *Enterprise*, along with nuclear escorts cruiser *Long Beach* and destroyer USS *Bainbridge* (CGN-25), ran square into Super Typhoon Faye, sending monstrous waves over the bow of *Enterprise* which towered 62 feet above the surface.

On 7 November the carrier arrived in Subic Bay for a three-week stopover. In addition to taking on supplies, making repairs, and flying practice missions, airmen attended JEST, taught by Negrito tribesmen of the Philippines. On 30 November, the day *Enterprise* departed Subic Bay en route to Dixie Station, Commander Sheldon Omar "Lefty" Schwartz arrived to the squadron as XO, replacing Commander Ron Andresen.

Schwartz entered the service as a NavCad in 1949, earning his wings on 1 June 1951 and subsequently serving in Korea. He had also served a 1956-1957 tour with the Blue Angels flying the Grumman F9F-8 Cougar, as well as an exchange tour with the United States Air Force at

Wright-Patterson AFB during the development stages of the F-111 Aardvark. Though Schwartz was not as personable nor affable as Commander Norman, he was nevertheless well-liked amongst the men in the squadron. Most important of all he was an excellent fighter pilot, a "good stick" as they would say, an aggressive aviator who would not hesitate to lead by example on the most dangerous and difficult missions.

Commander Bob Norman       Commander Sheldon Schwartz

Those in the squadron recall that it was not difficult to tell when Lefty was not happy with someone, or something. When necessary he would not hesitate to lay down the gauntlet to maintain discipline and order, particularly within the ranks of junior officers. He accomplished this in part by stern and gesticulating lectures in the ready room. This was the case when in a letter home a squadron officer disclosed to his wife some of the men's wild escapades while on liberty. She shared these details with several other wives, some of whom were not pleased with their husband's behavior, with the ensuing bickering threatening morale in the squadron. This resulted in a contentious ready room meeting where Lefty angrily proclaimed "What happens in this squadron during deployment…stays in this squadron during deployment!"

Men in VF-96 also recall that the Blue Angel in Schwartz would sometimes come out, particularly when flying in formation or returning to the carrier, where he would demand precise aviating and occasionally execute an aerobatic maneuver worthy of an air show. Coincidently there was also a Lieutenant (junior grade) Charles Schwartze within the squadron. To ease any confusion, Schwartze was also known as "Righty".

*Enterprise* began flying missions from Dixie Station on 2 December. Along with the nuggets who were flying combat for the first time, it was also the first time *Enterprise* had been in combat since being launched on 24 September 1960. For the nuggets in VF-96 and CVW-9 as a whole, they realized the skills they learned in training merely scratched the surface of what they would need to survive in combat. Doug Kindseth remembers:

> As RIOs, we trained on what the airplane was designed to do, intercept and shoot down enemy aircraft using missiles. We did not train on AAA, SAMs, or flak suppression, tasks not in the purview of the Phantom's mission. Yet, when we arrived in Southeast Asia that was all we were doing. There was a big difference between our training and the reality of combat. Most of our learning was on the job, and as a matter of survival, we quickly became competent in our trade.

Jim Stillinger recalls a flight he took with nugget pilot Ensign Lawrence "Steve" Amann. Veteran pilots in the squadron had warned Amann that when launching from the carrier with a full munitions load the control stick of the Phantom could not be as far aft as had been used during carrier qualifications. Stillinger recalls what occurred:

> We launched and Amann had the stick a little further forward than it should have been. As soon as we left the carrier we began a dive towards the water. It felt like our wheels were skimming the surface when I told him "Pull up, pull up Steve or I'm punching out!" He got it under control and we climbed up and away, making it a lesson I'm sure he did not soon forget.

Amann was far from the only VF-96 pilot experiencing challenges. Ritchie recalls a night landing on *Enterprise* following his first combat mission to South Vietnam:

> When we returned to the carrier the weather had completely changed and *Enterprise* was no longer where we expected it to be. We were up in the clouds and given radar vectors back to the boat. When I began my final approach for landing I could see the deck rocking, pitching up and down. It was unlike anything I had gone through in qualifications, where the seas were generally calm. I missed the wire on my first attempt and had to go around for another try. The task became physically and mentally arduous. I remember thinking to myself "This is serious business!"

Though Ritchie was ultimately able to land the Phantom on the pitching deck, his flight school classmate, also on his first combat deployment, was not as fortunate. After flying a strike on a target in South Vietnam pilot Lieutenant (junior grade) Robert Miller and RIO Lieutenant (junior grade) George "Duke" Martin returned to the carrier and found the same foul weather that Ritchie and others had encountered. They made several attempts to trap back on the carrier however they could not stabilize their Phantom (BuNo. 149468) for landing. Running low on gas, they attempted to rendezvous with an A-4 Skyhawk to refuel, however, weather conditions made it impossible. When the situation became critical and with insufficient fuel to bingo to Da Nang the two men had no choice but to eject alongside the carrier, and were rescued by a Kaman SH-2 Seasprite plane guard helicopter of HC-1 from *Enterprise*. Miller, a graduate of USNA, subsequently went to medical school and became a neurologist in the Navy Medical Corps. Following military service he entered private practice.

*Enterprise* had two F-4B squadrons within CVW-9, the other being sister squadron VF-92 call sign "Silver Kite". On that first day of line operations, the squadron lost their first aircraft of the conflict. Lieutenant T.J. Potter and RIO C.W. Schmidt call sign "Silver Kite 206" (BuNo. 151409) were providing close air support to American forces 5 miles north of An Loc in South Vietnam. Diving on the target, the men released their load of six Mk.82 bombs at 5,000 feet.

They immediately felt an explosion below their Phantom, with their wingman informing them they had a severe fuel leak and were trailing flames. With their gas tanks quickly emptying and fire engulfing the aircraft, they both ejected. Fortunately for them, a USAF C-123 Provider was in the area, who vectored in a USAF HH-43 Huskie helicopter to pick them up.

Flying CAP over the Tonkin Gulf.

Over the Mekong Delta.

It is theorized that one of the Mk.82s prematurely detonated due to a faulty electronic fuse, causing immense damage to the aircraft. The two men were incredibly lucky to have survived. After a stay of one week on Dixie Station the carrier moved 350 miles north to Yankee Station, arriving in mid-December. The targets of *Rolling Thunder* had expanded since the first deployment of VF-96, with roads, bridges, boats, and railroads as well as POL facilities appearing on the expanding target list.

The months of November-March in the north of Vietnam bring heavy rain and other inclement weather, which greatly affected the number of combat missions flown. Perhaps the foul weather was fortuitous, as another problem facing the air wing after arriving on Yankee Station was a shortage of Mk iron bombs, having to fill the deficit with WWII-era bombs of questionable reliability. These outdated and volatile bombs were later determined to have been a factor in the horrific fire aboard *Forrestal* on 29 July 1967. That conflagration, one of three carrier fires to occur on Yankee Station during the conflict, left 143 men dead and an additional 161 injured, many of them severely.

On 28 December, VF-96 pilot Lieutenant Dean Forsgren and RIO Lieutenant (junior grade) Robert "Deuce" Jewell found themselves in the same predicament as Miller and Martin two weeks prior. Returning from an armed recce mission the weather made landing on the pitching and rolling carrier extremely challenging. Running critically low on fuel and unable to tanker due to weather, they ejected from their F-4 (BuNo. 151438) and were rescued.

Navy aircraft from Yankee Station, including Phantoms, were in Laos regularly, interdicting the movement of military equipment and personnel from North Vietnam down the Ho Chi Minh Trail, where it wound in and out of both countries. According to the stipulations of the 1954 Geneva Accords that ended the first Indochina War, the United

States was not supposed to be operating in Laos. However, with Pathet Lao communist insurgents providing a haven for the trail and infiltration of arms and VC fighters into the south, there was little choice. Flying over Laos was a highly risky endeavor. If an American airman was shot down over North Vietnam and captured, there existed a chance of survival of the conflict as a POW. If shot down over Laos and captured the chances of survival were scant. Only two American airmen are known to have successfully escaped from Pathet Lao custody.

On the same night Forsgren and Jewell were forced to eject due to foul weather VF-92 lost a Phantom over central Laos while leading a *Steel Tiger* mission to an area near the Mu Gia Pass south of Vinh. Commander Edgar Rawsthorne, skipper of VF-92 and RIO Arthur Hill call sign "Silver Kite 203" (BuNo. 151412) had just fired a salvo of LAU-10 rockets at a target when their wingman witnessed the aircraft impact the side of a ridge near Ban Pondong. No enemy fire was observed, leading to the conclusion that Rawsthorne became spatially disoriented or failed to recover from his dive in time to clear the terrain. Both men were ultimately declared KIA with neither body recovered.

One week later, on 18 February, VF-92 suffered another loss. Lieutenant (junior grade) James Ruffin and RIO Lieutenant (junior grade) Larry Spencer call sign "Silver Kite 201" (BuNo. 152297) were escorting a Lockheed EC-121 Warning Star on a surveillance mission off the coast near Thanh Hoa. The EC-121, equipped with massive radomes on the top and belly of the fuselage, was the military version of the famous Constellation airliner. Without warning a massive explosion, suspected to be a SAM, mortally damaged the Phantom. The aircraft immediately pitched down in a dive toward the water. Both men ejected, however only Spencer returned during *Operation Homecoming*. Ruffin's remains were repatriated in June 1983.

In a country filled with rivers, tributaries, and other waterways, rail and vehicle bridges were common targets. Most notably, the Paul Doumer Bridge over the Red River in Hanoi and Dragon's Jaw Bridge over the Song Ma River near Thanh Hoa were early Alpha strike targets of *Rolling Thunder*. Doug Kindseth shares the frustration felt by many airmen who attacked these heavily defended bridges numerous times, seemingly with little effect:

> We would bomb a bridge out of commission. The next time we flew to the area, it would be back in service! They could repair bridges at astonishing speed, and of course, some bridges were attacked unsuccessfully time and time again. This was certainly the case with the concrete and steel Thanh Hoa Bridge.

Ritchie echoes Kindseth's sentiments when recalling several challenging missions to the Hai Duong Bridge, located within the Iron Triangle:

> Every time we attacked that bridge we were sure it had been destroyed. However, when the dust settled and intelligence came in the bridge would still be standing. Whatever damage we wrought the North Vietnamese would quickly repair. Like other major bridges in North Vietnam, we attacked it many times, losing numerous men and aircraft. It was both frustrating and maddening.

On 20 March a mission to a North Vietnam bridge claimed another aircraft when VF-92 lost their fourth Phantom of the cruise. Lieutenant (junior grade) James Greenwood and RIO Lieutenant (junior grade) Richard Ratzlaff call sign "Silver Kite 202" were attacking a bridge ten miles southwest of Vinh when their Phantom (BuNo. 151410) was badly damaged, likely by shrapnel from their own bombs. They turned east and had just made the coast when their aircraft became uncontrollable, forcing them to eject north of Mu Ron Ma. They landed in the water several hundred yards from the coast, which had numerous AAA guns. Ratzlaff was captured by a North Vietnamese fishing boat. Greenwood swam out to sea, and after a short but intense battle with enemy coastal forces involving a USAF HU-16 Albatross amphibian, Navy Seasprite helicopter, and eight 20mm cannons of four A-4 Skyhawks from *Enterprise*, he was rescued. Ratzlaff was later released during *Operation Homecoming*.

By the time the 1965-1966 WESTPAC concluded in June *Enterprise* had spent 131 days on the line. The *Fighting Falcons* of VF-96 had flown over 2,000 hours in combat, flying 1,250 missions while dropping more than 700 tons of ordnance. Skipper Norman had stayed true to his word, with VF-96 not suffering a single casualty that deployment, despite high losses by sister squadron VF-92. Many in the squadron attribute this to the leadership of not only Norman but also Schwartz and other veteran officers within. After the cruise *Enterprise* did not return to Norfolk. It instead steamed to NAS Alameda, where it would henceforth be based.

After returning to NAS Miramar Commander Norman was relieved by Lefty Schwartz, with Lieutenant Commander Richard Rich arriving to the squadron as XO. On his first combat deployment, Rich was a quiet and reserved Navy officer; meticulous and by-the-book. Unlike

Commander Norman or even Lefty Schwartz, few men in the squadron had what could be considered a personal relationship with him. Many in VF-96 feel that Rich was "checking boxes" and on the fast track to a command position, most likely as their next skipper. The squadron spent the remainder of the summer and early fall under the tutelage of their new skipper and XO, relentlessly preparing for their upcoming November departure. Pilots and RIOs recall that both Rich and Schwartz were strong proponents of aerial combat training, stressing the many lessons that had been learned since the first MiGs were encountered over North Vietnam during the opening stages of *Rolling Thunder*. Thus, much of their focus that summer was on dogfighting tactics. This was accomplished at Miramar, as well as the Yuma and El Centro ranges.

The presence of MiGs exposed a rare but crucial deficiency in the Phantom. Designed in a pre-ICBM world for a defensive role against enemy aircraft, primarily Soviet nuclear-equipped bombers, the Phantom's mission was to intercept, engage and destroy these threats. With the advent of "fire and forget" missiles such as the AIM-9 Sidewinder and AIM-7 Sparrow, conventional wisdom at the time was that the era of the dogfight was over, and that future aerial combat lay in missiles. Thus, the four internal 20mm cannons originally proposed by McDonnell were eliminated during the design phase, deemed obsolete and unnecessary. On the other hand, MiGs had powerful internal cannons. They also carried K-13 Atoll air-to-air missiles, a Soviet example of a reverse-engineered Sidewinder.

Early aerial engagements in the conflict showed the decision to remove the cannons to be deeply flawed. Air-to-air missiles could only be used within a specific envelope. They could not be used at very close range and required a lock on the target before they could be fired. They could not be used BVR as ROEs required the pilot to visually identify a target before firing. If a Phantom did fire a Sidewinder or Sparrow, statistics showed limited chances of success. During *Operation Rolling Thunder* disappointing kill rates of 18% for the AIM-9 and 9.2% for the AIM-7 left many pilots with little confidence in the weapons.

Though the F-4 models flown by the Navy did not have internal cannons, gun pods could be mounted on either the center or wing stations. However, these pods produced a series of their own problems. They were heavy and bulky, resulting in increased drag and decreased performance. They were also prone to jamming after only a few bursts. Lastly, they were meant primarily for ground versus air targets.

The lack of an internal cannon and unreliable missiles was not the only problem confronting Phantoms in the Vietnam theatre. Close engagements with MiGs had shown that they could out-maneuver a Phantom in a horizontal fight, particularly at low altitudes and speeds. They were half the weight, with a smaller, nimbler profile and a tighter turn radius. However, the Phantom was faster and more powerful than any MiG of the time and had much better climb performance. In addition, the Phantom could carry up to eight air-to-air missiles, while the MiG-21 was limited to two Atolls.

A favorable tactic was using the power of the F-4 to climb vertically, then use speed and altitude to maneuver into a position to fire a missile. Though dogfights with MiGs did occur, in reality, VPAF pilots did not prefer to engage in close combat with American aircraft. Instead, their favored tactic was to ambush, fire their weapons, and run.

After careful study of the deficiencies, in 1968 the Navy released the Ault Report which cited numerous problems. Though there were issues with the missiles themselves, a lack of adequate training was also cited as a major factor. This led, in 1969, to the creation of NFWS, more popularly known as "Top Gun". Training squadron RAG VF-121 at NAS Miramar was chosen as the location of NFWS. In addition, the Air Combat Maneuvering Range (ACMR) at MCAS Yuma was created to further train in the art of dogfighting. As time went on issues with the missiles themselves were addressed, and along with new aerial combat training curriculum, remedied many of the problems. The Air Force took a direct approach, adding an internal M61 20mm cannon to the F-4E model, which was not used by the Navy. The first USAF F-4Es reached the Vietnam theatre in November of 1968. The "E" model was manufactured in greater numbers than any other model of the Phantom.

Very few endeavors in life forge close bonds between people than serving in the military together, particularly in combat. This was certainly the case with VF-96, who were a very close group, professionally, and personally. Aside from spending months on end shoulder-to-shoulder, when back at NAS Miramar they also spent many hours away from the flight line in each other's company. The carousing would begin on Friday evening with happy hour, which would segue into Dance Night at the Marine Corps Recruit Depot (MCRD) O-Club in San Diego. Closing out the weekend was Sunday night at *Downwinds*, a club at NAS North Island featuring the musical grooves of local San Diego band Linda and the Centaurs. These libatious indulgences were more than exercises of traditional military culture. They also served both as a bonding and stress relief mechanism; a time when they could simply enjoy their camaraderie before once again putting their lives on the line.

As summer marched on and with fall rapidly approaching, the squadron's November deployment loomed on the horizon. The zeal that was felt by many aviators when first deployed to SEA had largely dissipated, replaced by cynicism in the manner in which Washington, DC was conducting the effort, the micro-managing of the military by the Johnson Administration, and the staid reality of a bloody conflict that was quickly spiraling out of control. The loss of lives and aircraft was becoming a near-daily event with each and every one felt throughout the Naval Aviation community. Jim Ritchie remarks:

> Much of what we knew of the situation in Vietnam came from the government, which had been simplistically portraying the conflict as good versus evil, freedom versus communism. Of course, there was some truth to that, however as the conflict dragged on and losses mounted, we became much more cynical. We were out there flying, fighting, and losing many fine men—husbands, fathers, and sons. Yet, none of the sacrifices seemed to be making much difference. It just droned on and on, with no real progress or end in sight.

Dave Hoffman shares these sentiments:

> I recall sitting in the ready room with other members of the squadron, looking at the targets. We were scratching our heads as many of them made no sense and had little to no impact. Not only that, but we were going to these insignificant targets over and over, losing men for what seemed like no good reason. However, our duty was to follow orders and fly missions we were assigned, regardless of our personal feelings, and that's what we did.

VF-96 *Fighting Falcons* 1966-1967 *Enterprise* WESTPAC

Front Row: Schumacher, Engelmann, Boehmer, Frank, Daley, Galler, Skinner, Stark, Nordell, Markey, Roberts, Wohlfiel, Ewing, Carrell, Stillinger, Kindseth, Liscum, Hollarn Back Row: Ritchie, Amann, Winchester, Hill, Earnest, Schwarze, Beasley, Rich, Schwartz, Gloves, Wear, Forsgren, Hoffman, Wilson, Baldry, Dwyer, Born, Welch, Wagner

It should be noted that no airman is forced to fly—they volunteer for this hazardous duty. At any time they can turn in their highly coveted wings, spending the remainder of their commitment in relative safety. However, historically this has been exceedingly rare. This held true in the Vietnam conflict, even as the odds rose against them and losses markedly increased. It should also be noted that it was becoming common knowledge within the military that the North Vietnamese were torturing,[47] and in some cases murdering American POWs, while the majority of unfortunate souls who went down in Laos were disappearing without a trace.

By the fall of 1966, the frenzied pace of the conflict was being reported on extensively by American news organizations. Newspapers such as the *San Diego Union-Tribune* were publishing daily ever-growing casualty lists. No longer a footnote of casual mention, the rapidly growing conflict, along with vehement anti-war sentiments emerging in the country, made the conflict prominent, front-page news across the United States. Though many Americans had become generally familiar of the events occurring half a world away, the military was watching very closely, with increased concern and consternation. This was not exclusive to the uniformed ranks. On the evening of 18 November Doug Kindseth, Jim Stillinger and Duff Cooke were having a farewell dinner with their wives at *The Elegant Farmer* restaurant in Oakland. Kindseth and Stillinger were roommates on *Enterprise* and at the time joked that they spent more nights together than they did with their wives. As the live entertainment began a rendition of "I Left My Heart in San Francisco" the wives broke down and began crying. The men were hard-pressed to give consolation, themselves drawn into a web of emotions on the eve of their deployment.

The following day VF-96 departed NAS Alameda aboard *Enterprise*, the squadron's third deployment to Vietnam. After steaming from the west coast the carrier made customary five-day stops in Pearl Harbor for ORI, followed by Subic Bay. On 15 December the carrier departed Subic and steamed straight to Yankee Station, foregoing acclimation on Dixie Station. The air wing began flying combat sorties over North Vietnam on 18 December, with an expanded target list of industrial facilities, power generating plants, and other critical infrastructure. This targeting would quickly expand to include, for the first time, objectives within the heavily defended cities of Hanoi and Haiphong.

A new addition to VF-96 for the 1966-1967 cruise was Lieutenant Terry Born, a highly experienced and aggressive pilot who had flown from *Constellation* with VF-143 during *Pierce Arrow*, providing CAP during the strike on the POL storage facility at Vinh. A second deployment with VF-143 aboard *Ranger* January-August 1966 followed. He arrived to VF-96 in September after volunteering for a third combat deployment to Southeast Asia, just before the squadron's November departure. For most, two combat deployments would be quite enough. However, for Born, his enjoyment of the risks and challenges outweighed any inherent dangers:

> Can you believe the Navy was paying me to fly an F-4? I probably would have done it for free! At the time I wasn't married, so what else would I have been doing? Piloting airliners? I wanted to fly combat!

Born's first RIO on that cruise was Ensign Joe Galler, who left the squadron in early March to attend flight training. A young and inexperienced nugget RIO, Ensign Garrie Liscum, fresh from RAG

VF-121, replaced Galler. In a practice highlighted previously Liscum, the most junior RIO in the squadron, was paired with Born, who had more combat hours in an F-4 than any other pilot in VF-96.

He well remembers his first mission with the nugget RIO. As they were diving on the target Liscum called for bomb release. Afterward, Terry realized their munitions had not even come close to the target. When he pointed this out Liscum's truthful response was "Well, in the RAG we didn't practice with people shooting at us and smoke from explosions filling the air!" In short time, under Born's tutelage, Liscum markedly improved and became a skilled RIO.

As with previous winter cruises, weather greatly curtailed combat missions over North Vietnam, with many scheduled strikes canceled due to cloud cover and visibility. However, on 4 February weather had improved enough for Alpha strikes against important targets in and around Thanh Hoa including rail yards, sidings, and spurs as well as bridges, roads, and storage facilities. A total of nine daylight strikes were launched that day from *Enterprise*, *Kitty Hawk*, and *Ticonderoga*.

The attack aircraft utilized by the Navy for the majority of Alpha strike missions was the subsonic, single-engine A-4 Skyhawk. Small, lightweight, and agile, the Skyhawk was the de facto Navy aircraft fulfilling the light attack role in Vietnam. However, by 1967 the air defenses in North Vietnam had improved to the point where A-4s were extremely vulnerable to AAA and SAMs. When fully laden with bombs and other munitions the normally agile Skyhawks, particularly the "C" models, became slow, easy targets. Therefore Alpha strikes into North Vietnam typically had the F-4 in a flak suppression role, though they occasionally attacked Alpha strike targets as well.

Russ Goodman, Duff Cooke, Lefty Schwartz, Jay Henglemann (back to camera) and
Dick Rich enjoy a feast on Christmas Day 1966.

Jay Henglemann, Dave Hoffman, Russ Goodman and Jim Stillinger celebrate
Christmas Day 1966 in the ready room.

251

Strike force flak-suppression Phantoms would launch from the carrier first and fly ahead of the main strike group, targeting air defenses. As they dove on the AAA sites they were confronted with muzzle flashes of guns, streaking tracers, explosions of flak shells, and often SAMs. At times the flak bursts were so heavy it seemed like they could walk to their targets on the deceivingly deadly, puffy black clouds.

The most common AAA calibers encountered in the theatre were 23mm, 37mm, and 57mm. However, AAA calibers of 85mm and 100mm were also present in the north, particularly around Hanoi and Haiphong. These AAA guns were often radar-guided, and aircraft hit by these larger calibers rarely stayed in one piece, or in the air for very long after. Kindseth, who flew with Lieutenant Commander Bill Dwyer on the 1966-1967 cruise, remembers the high amount of danger:

> When those larger calibers exploded, we saw and felt it. We would get bounced around pretty good in the airplane when one of those went off anywhere close to us. I know we never took a direct hit from one, as if we did I would likely not be here describing it.

Although crews flying flak suppression roles were extensively briefed on air defenses before embarking on a mission, including types, locations, and often recon photographs, there were occasions when Phantoms arrived over the target to find air defenses shuffled and in different locations, particularly SAMs. Terry Born recalls:

A recon bird would fly over a target and take pictures of air defenses, which we would often see in our pre-flight briefings. However, the North Vietnamese were smart and often moved the defenses around, particularly at night. They also created dummy revetments where it only looked like a AAA site. If AAA was not where we expected it to be, it was usually not too difficult to find it, or another target.

With attack aircraft launched and on their way, the Phantoms would have to scramble to locate air defenses and take them out before they arrived, all while deftly maneuvering to avoid being shot down themselves.

Jim Hollarn in IOIC working on identifying aircraft.

Lieutenant Jim Hollarn arrived to VF-96 on 14 April 1966 as an AIO, meeting *Enterprise* in Subic Bay. He specialized in AAA and SAM intelligence, working in the carrier's IOIC. His job was to disseminate and analyze intelligence from not only CVW-9 but from other air wings on Yankee Station as well. After reviewing and correlating AAA and SAM intelligence he would mark the sites on a large map in IOIC using different colored pins and stickers to denote differing calibers of AAA and SAM sites. Hollarn would brief crews on types and locations of air defenses before a mission, and debrief them after. He recalls the fluidity of North Vietnamese air defenses and the difficulties that presented when planning missions within the country:

> Cities such as Thanh Hoa and Vinh were target-rich environments, and therefore heavily defended. There were not many ingress and egress routes that didn't have some degree of risk. We would try to stay one step ahead of the North Vietnamese, however, they were very clever and often kept pace with our efforts. It was a high-stakes cat-and-mouse game.

The AAA gun emplacements that defended Alpha strike targets were often revetted, and protected by concrete, sandbags, protective metal sheets, and other reinforcements. The Mk.82 bombs used for flak suppression were generally proximity-fused to detonate above the ground, showering shrapnel down on the AAA, and bypassing many of the lateral reinforcements. In the latter part of the cruise, the squadron began carrying CBU cluster bombs, which were extremely effective against revetted guns. Kindseth recalls the rampant destruction wrought by cluster bombs on AAA sites:

When a cluster bomb detonated it was hundreds of small bombs ("bomblets") exploding. Afterward, the entire area looked like a hurricane had hit. Complete and utter destruction, often leaving the site permanently disabled and devoid of any life.

After performing flak suppression Phantoms would assume a CAP role, protecting the Skyhawks, Intruders, and other attack aircraft during the strike. To guard against airborne threats the F-4s would also be armed with air-to-air missiles. The Phantoms providing flak suppression and CAP would be the first aircraft over the target, and the last aircraft to depart the area following an attack. Their time over the target from arrival to egress was typically 15 minutes.

In addition to large AAA guns, revetted and otherwise, aircraft also faced threats from small-caliber AAA such as long rifles. Many a Phantom and other costly aircraft were brought down by small arms, by damage to the aircraft, or incapacitation of a crew member. It was not at all unusual for reconnaissance photographs of North Vietnam and Laos to show shadowy figures standing in rice paddies or fields, holding firearms of various calibers, pointed toward the sky. Kindseth remarks:

> With a large number of armed soldiers and civilians in North Vietnam and Laos, it was likely that every minute we were over the countries somebody had us in their sights. We therefore learned very quickly that the threat was ever-present. You paid attention all the time and did not make mistakes. Guys that did not pay attention or made mistakes often paid a price, with heavy consequences.

RIO Ensign Gary Thornton, a nugget who joined VF-96 for the 1966-1967 cruise, echoes Kindseth's sentiments:

> From observation, I can guarantee that if we were within a mile of the NVN shoreline someone was shooting at us! All the locals had weapons, and they knew how to use them. Even on seemingly benign flights that were not classified or counted as combat missions, we would have people on the ground firing at us. We were like clay targets at a trap range.

Although statistically AAA was the greatest danger to aircraft over North Vietnam, SAMs were also a formidable threat, with their numbers steadily increasing over the course of the conflict. By 1966 many important targets in the north, as well as ingress and egress points for aircraft on Yankee Station, were defended by SAM batteries. The most effective strategy for eliminating SAMs was to destroy them on the ground before they were fired. This was accomplished by SAM suppression either with conventional weapons or AGM-45 Shrike missiles employed by Navy "Iron Hand" aircraft. The Air Force also had specialized SAM-hunting "Wild Weasel" squadrons. These units proved to be very effective at locating, and destroying, SAM sites.

As Alpha strikes were meticulously planned down to the second, the success of the mission depended on dozens of aircraft with different roles in the strike being at the right place, at the right time. The North Vietnamese would often fire a volley of SAMs at a strike group, putting numerous missiles in the air, forcing evasive action. This would also cause aircraft to fly lower, opening them up to AAA.

In the latter part of the 1966-1967 cruise VF-96 Phantoms began receiving AN/APR RHAW receivers to alert crewmen to radar-guided AAA and SAM activity. Along with audible warnings in their headsets, the system also included a small scope indicating the bearing of the threat. However, crews found that having RHAW active while over North Vietnam resulted in non-stop alerts due to being constantly swept by enemy SAM and AAA radar. Consequentially, many times the RHAW alerts were muted to eliminate the annoying distraction.

When a SAM locked onto an aircraft the pilot had several options at his disposal. One such option was a well-timed, high-speed, highly aggressive maneuver that could break the radar lock of a SAM and eliminate it as a threat. The SA-2 SAM was thirty-five feet long and traveled in excess of Mach 3. If a pilot was able to track the SAM from launch he would maneuver to put it at an oblique angle to the nose of the aircraft. When the missile was mere seconds from impact the pilot would execute a sharp turn or diving bank toward the missile's trajectory. Traveling at supersonic speed the SAM could not turn to maintain a radar lock on the target and would subsequently detonate. If flying above a cloud layer with no ground visibility, the pilot would have little to no time to react to a SAM piercing the undercast cloud deck, and thus have a reduced chance of evading the missile.

If a pilot managed to out-maneuver a SAM he was still not necessarily out of danger. With a 430lb. fragmentation warhead and blast radius of up to 820 feet, the aircraft would have to be beyond this distance to have a chance at escaping. Playing chicken with a telephone pole traveling at supersonic speed with nearly a quarter-ton of explosives in a fragmenting shrapnel warhead was perilous business. It required superb airmanship and sharp reflexes, along with precise timing and a healthy dose of boldness and courage. There was little room for error.

Firing rockets on a target in North Vietnam.

While this WESTPAC would see a marked increase in Alpha strikes, armed recce missions continued to compose a majority of the missions. These targets were usually trucks, barges, sampans, junks, or other motorized vehicles used to move personnel and supplies. Though many of these targets ended up on the receiving end of American ordnance, the North Vietnamese were masters at concealing their movements, leaving many to complete their journeys south under the cover of jungle canopy and darkness. Terry Born explains:

> Night armed recce missions were great for us. Much of it was bombing trucks that were moving south, so we would typically fly north along roads like Route 1. Nighttime made it easier to spot the trucks. Headlights were often darkened, but you could still see them. Some of the night recce missions were "Blind Bat" where parachute flares would be dropped from A-3s, C-130s, or other aircraft, making it easier to see the targets we were attacking.

Though the flares were useful for us, they were also useful for the enemy. That telegraphed to them that an attack was underway, and they would use the illumination of the flares to conceal their vehicles under the jungle canopy. My preference on the night road recce missions was to come in under darkness and lay a string of bombs up a road, south to north. By the time the enemy looked up the bombs were already on their way. We would roll 90° or so as we pulled up from the run. Many good hits were observed with lots of large explosions.

RIO Jim Stillinger recalls a particular night road recce mission over North Vietnam He was flying with pilot Lieutenant Commander Ken Baldry, on the hunt for trucks and other vehicles. They were being assisted by an RA-3B Skywarrior "Q Whale"[48] equipped with IR cameras to spot heat signatures of vehicle engines in the darkness. Baldry and Stillinger's Phantom was carrying a load of Mk.77 bombs, while their wingman was carrying illumination flares. He recalls:

The Whale spotted numerous heat signatures in the darkness, so our wingman dropped flares to illuminate the area. We spotted something, but it was difficult to tell what in the eerie light the flares produced. The RA-3B insisted there was something, so we dropped our bombs, resulting in numerous tremendous explosions. At that time a group of A-4s appeared and asked if they could lend a hand. They dropped their munitions, which resulted in additional explosions.

The Skywarrior, along with the Phantoms, had uncovered a large arms convoy moving south under cover of jungle canopy and darkness. The IR sensor had betrayed their location and any attempts at concealment, resulting in their destruction. For Dave Hoffman, night recce missions, particularly in the mountainous areas north of Hanoi and Haiphong, were the most challenging, and frightful:

> Some of those nights were pitch black, and outside of the cities, the terrain was just as dark as the night sky. We often didn't have any idea what was below us in terms of air defenses and AAA would appear out of nowhere. SAMs were particularly threatening in the darkness. If SAMs appeared at night we would often dump whatever we had on the stations, get low and feet wet as quickly as possible. Trying to evade SAMs in the darkness of night above mountainous terrain was extremely risky.

On daytime recce missions, the Phantoms would go to a location of known enemy activity such as a river crossing. Occasionally they would see pontoon bridges, waiting to be reassembled at night. In the absence of targets, the Phantoms would drop their bombs on roads, cratering them, thus making them unusable. However, once the sun set the North Vietnamese would quickly repair the damage.

Terry Born recalls a 1 March daytime recce mission with RIO Galler, accompanied by Jim Ritchie and RIO Schumacher as wingmen. They were patrolling a road near Cam Pha, northeast of Haiphong when they spotted a suspicious truck speeding along. As Born commenced his attack on the target Ritchie called over the radio "They're shooting at you, but it's behind you".

Windscreen damage from AAA on F-4 flown by Terry Born and Joe Galler
1 March 1967 (tech is in pilot seat facing rear).

Just as Born was about to release his Mk.82 Snake Eye bombs a AAA slug impacted the windscreen. Unfazed, he continued the attack, dropping his ordnance. After pulling up he glanced in his rear-view mirror to check on his RIO. Galler had a look of shock on his face, and as Born describes it "His eyes were as big as silver dollars!"

"What happened? What was that?" a stunned and concerned Galler inquired. Born calmly replied "You can see it. Look up here!" To reassure himself and his RIO they were not in any danger he scanned his instruments, which showed that the Phantom was perfectly airworthy.

The projectile had opened a gash on the left side of the windscreen several inches long. Along with leaving a red-colored residue in the cockpit, fragments from the shell and windscreen sprayed Born, who received small cuts on his face as a result. Had the projectile completed its course as intended by the person who fired it, Born and quite possibly Galler would not have survived the mission.

After learning of the close call wingman Ritchie suggested the two aircraft not press their luck, and return to *Enterprise*. Unflappable flight leader Born rejected the suggestion, and their sortie continued, eventually finding barges for their remaining ordnance. Strictly by coincidence, this was Galler's last mission with VF-96. He departed for pilot training shortly thereafter, with Ensign Garrie Liscum taking his place. To this day Born still jests Ritchie over the "They're shooting at you, but it's behind you" remark, issued seconds before he and Galler were hit.

There was one type of mission that Phantom crews typically enjoyed regardless of the risk—escorting RA-5C Vigilante aircraft on their reconnaissance flights. As the Vigi was unarmed they were required to have an F-4 escort with them when flying over North Vietnam. While the Vigis gathered intelligence the Phantoms monitored for MiGs and kept on the lookout for AAA and other threats.

The Phantom and Vigi shared the same powerful General Electric J79 after-burning engines and were both exceptionally fast. In fact, at the time they were the two fastest aircraft deployed on carriers. However, the Phantom was limited to 600 knots when carrying external stores. The Vigi, with a clean airframe, had no such limitation. Often unable to keep up while on their supersonic runs, the F-4 would hold at a predetermined point while the Vigi gathered intelligence, rendezvousing afterward.

Terry Born recalls his first time escorting a Vigilante, on a mission to Kep Airfield, the large VPAF airbase located 30 miles northeast of Hanoi. He remembers well the challenge of keeping pace with the slick RA-5C:

> We had to stay in full afterburner, maximum power, just to try and keep up with the Vigi, which we couldn't. By the time we reached Kep and flew over the field, the Vigi crew was already done with their photo run and were completing a 180° turn to get feet wet on their way back to *Ranger*.

Though escorting a Vigilante was more often than not routine and uneventful, on 12 February that changed for Major Russ Goodman, an exchange pilot from the Air Force who had served a 1964-1965 tour with the Thunderbirds aerobatic team, and his RIO Gary Thornton. They were escorting a Vigilante of RVAH-7 call sign "Flare 105" from *Enterprise*. Commander Don Jarvis and RAN Lieutenant (junior grade) Paul Artlip were on a coastal SLAR mission when they were shot down by AAA 30 miles northeast of Thanh Hoa. Thornton remembers the incident well:

> We were a few hundred yards behind them, at their four o'clock position. We saw 85mm flak go through their outboard starboard wing. Russ and I climbed up through the overcast, but the Vigi didn't. We then dropped back down and began our search. After a couple of racetrack laps around the area I spotted a flat slick in the water in

the shape of an RA-5C. Another couple of laps and we spotted the crew in the water, just offshore.

With word of the situation spreading Goodman and Thornton were soon joined by an E-2A Hawkeye, two SH-3 Sea Kings, four A-1 Skyraiders, and a USAF HU-16 Albatross amphibian along with four additional Phantoms. Several of these aircraft spotted a North Vietnamese motorized junk that heading toward Jarvis and Artlip.

These motorized junks were constructed mostly from wood, making it difficult to get a radar sweep or acquire a lock with a radar-guided Sparrow missile, an air-to-air weapon. However on an inbound track towards the men Thornton was able to sweep the junk on his radar. He asked Goodman to extend his next inbound track by a mile to give him time to acquire a lock. On the next inbound track, Thornton was able to lock onto the boat, and at a distance of about two miles, they fired a Sparrow missile. Too low to properly track the target, the missile splashed into the water a mile from the boat. However, the message was received by the vessel, which quickly left the scene. Jarvis was picked up by the HU-16, while Artlip was plucked from the water by an SH-3.

That night pilot Lieutenant Commander Martin "Marty" Sullivan, who was the maintenance officer of VF-96, and RIO Lieutenant (junior grade) Paul Carlson were flying training exercises over the Tonkin Gulf with Skipper Schwartz and RIO Lieutenant Commander Dean Nordell. On the third exercise, their Phantom (BuNo. 152219) descended steeply into an undercast cloud layer and was not seen again. The last communication from the two men was a click on the radio followed by silence. The ensuing search included the carrier USS *Bennington* (CVA-20) and destroyer escort USS *Bauer* (DE-1025). All that was found was an oil slick and a small amount of debris on the water's surface.

Flying CAP with air-to-air missiles ready.

Art Cisson manning up for a mission.

It is suspected that the dark night with no visible horizon resulted in spatial disorientation. Sullivan and Carlson were the first VF-96 fatalities since Murphy and Fegan on 9 April 1965.

Eight days later, on 20 February, VF-96 launched two F-4s on a mission to attack a railway siding at Thien Lin Dong, eight miles southwest of Thanh Hoa and eighty miles south of Hanoi. Leading the flight were Goodman and RIO Thornton call sign "Showtime 614" (BuNo. 150413). Their wingman was Lieutenant (junior grade) Charlie Hill and RIO Lieutenant (junior grade) André Frank call sign "Showtime 615". Thornton and Frank were friends as well as roommates on *Enterprise*.

Approaching the target, Goodman climbed to 11,000 feet and commenced a dive. As the Phantom screamed toward the target muzzle flashes from AAA ground fire erupted like flash bulbs. Tracers and bursts of black streaked by the aircraft.

Goodman had commenced his dive at an altitude 1,000 feet higher than normal, therefore his dive angle was steeper than the standard 45°. This caused some difficulty in getting the aircraft lined up correctly to release the load of Mk.82 bombs. By the time he was satisfied with the orientation the aircraft was at 4,000 feet, a thousand feet lower than the standard altitude for bomb release. After pickling the bombs Goodman began a 5g pull-out. The nose of the aircraft was just about to clear the horizon when a AAA shell exploded just outside the port engine intake. Thornton vividly remembers what happened next:

> I saw and heard the AAA shell detonate The left engine ingested some of the shrapnel, which caused a catastrophic failure resulting in an explosion. The right engine then exploded.

Those explosions must have ruptured our fuel cells because years later our wingman Charlie Hill told me it looked like the entire aircraft had disintegrated into a fireball.

After regaining his senses, Thornton realized that Goodman had been hit and was no longer able to pilot the aircraft:

> I believe that when that shell exploded some of the shrapnel penetrated the fuselage or windscreen, entered the cockpit, and hit Russ. The reason is that after we were hit we started to porpoise and roll side to side. I think he still had the control stick in his hand because as we would roll Russ would lean in the opposite direction. This happened three or four times. During this period the nose of the aircraft began to drop lower on the horizon, altitude began to decrease while airspeed began to increase. I knew it was time to get out.

Now only a few hundred feet from the ground, feet dry and mere seconds away from impact, Thornton reached up and pulled the face curtain ejection handle above him position, which fired the canopy into the sky and raised his seat. The ensuing gale-force winds entering the cockpit blew his hand off the handle before he could pull it to the second position to complete the ejection cycle. He managed to grab ahold of it again, pulling it hard which ignited the explosive charge[49] he was sitting on, propelling him up and away from the stricken Phantom. Thornton remembers that he had perhaps ten seconds and one or two swings in the parachute before his feet hit the ground, and less than a minute after that before he was captured. Goodman, likely seriously wounded by AAA, never exited the airplane. Crash site excavations between 1993 and 2008 yielded his remains.

Wingman Hill and RIO Frank, who had just commenced their dive on the target when Goodman and Thornton were hit, broke off the attack. They flew eight miles east to get feet wet, dropping their ordnance into the Tonkin Gulf. They then returned to the area to assist in a rescue, however, there was no communication or visual on either man. This absence of contact, along with the fiery explosions, subsequently led to an assumption by their squadron mates that both men had been killed. The following day Doug Kindseth's wife clipped an article from the *San Diego Union-Tribune* for the scrapbook she was assembling for her husband. It was titled "REDS SHOOT DOWN, KILL FORMER JET THUNDERBIRD".

On the night of 25 February, after catapulting from *Enterprise* pilot Lieutenant Dave Hoffman and RIO Lieutenant (junior grade) Robert Ewing experienced an unrecoverable engine failure in their Phantom (BuNo. 152989). Hoffman remembers that a half-hour after departing the carrier and 50 miles north of the boat he received an engine fire warning light. After shutting down that engine and flying back to within 15 miles of *Enterprise* the other engine failed, leaving him few options. Both men successfully ejected and were quickly rescued by the carrier's plane guard helicopter.

Nearly five years later Hoffman would again eject from an F-4, after flying off *Coral Sea* as part of VF-111. On 30 December 1971, Hoffman and RIO Lieutenant (junior grade) Norris Charles call sign "Old Nick 203" (BuNo. 150418) were flying a MiG sweep near the city of Vinh, clearing the skies for the Air Force to come in and bomb a transshipment point. While orbiting over the city at 25,000 feet they dodged four SA-2 missiles that emerged from the undercast cloud layer below. However a fifth escaped their efforts and detonated near their Phantom, blowing the empennage off. With the aircraft in a steep, unrecoverable dive the two men were forced to eject feet dry and were subsequently captured. Hoffman, who was an LSO for CVW-15, was on his 205th combat mission. Charles accepted early release in September 1972. Hoffman was released in early 1973 during *Operation Homecoming*.

As winter faded weather began to improve over the north. The number of Alpha strikes increased, with some in the air wing flying multiple combat sorties every day. Strikes in and around Hanoi and Haiphong also increased. These cities and infrastructure were critical to the war machine of North Vietnam. As such they were heavily protected by SAMs and rings of deadly AAA guns.

On 4 April sister squadron VF-92 suffered their only losses of the cruise when two Silver Kite F-4s collided while on a BARCAP mission. Both RIOs—Lieutenant Philip Szeyller (BuNo. 152984) and RIO Ensign David Martin (BuNo. 151493) lost their lives in the mishap. Pilots Commander J.L. Rough and Lieutenant (junior grade) C.R. Jones ejected and were rescued.

On the morning of 8 April Jim Ritchie and RIO Frank Schumacher call sign "Showtime 610" (BuNo. 152978) along with wingman Lieutenant Commander Bill Dwyer and RIO Doug Kindseth call sign "Showtime 611" were flying north over the Tonkin Gulf on an armed recce mission to the Haiphong area. They were feet wet 15 miles east of Hon Gai when the weather began deteriorating, prompting the two Phantoms to turn back to *Enterprise*. Ritchie spotted a small island ahead, shrouded in haze. Knowing there were air defenses on these coastal islands he made a hard bank to avoid flying over it. He had just begun to turn when he felt and heard a distinct thump. Believing the source to be his RIO stowing the radar, Ritchie remarked to not stow it when near land. Schumacher, in a voice tinged with concern, replied "I didn't stow the radar".

The first sign was the illumination of several warning lights in the cockpit, quickly followed by a loss of avionics and hydraulics. Ritchie glanced in the rear-view mirror on the left-hand side of his cockpit and saw Schumacher in the back seat, wide-eyed and staring at him intently. He pitched the nose up, trying to gain some altitude that would buy them additional glide to get further away from the coast and out to sea, where their chances of rescue were better. However, the Phantom continued to pitch up beyond Ritchie's control inputs, leaving him little choice. He jettisoned his canopy, a pre-arranged signal to Schumacher in the back seat to punch out, which he did. With the aircraft now vertical and airspeed decaying, Ritchie followed suit. In an instant, he was out of the Phantom and floating down to the water. He spotted Schumacher descending about 1,000 feet below him, and their burning F-4 diving towards the gulf. of the memorable incident he remarks:

> Ejecting was a very surreal experience, realizing somebody on the ground had shot you out of the sky. One second everything was routine and the next I was wrestling with a 40,000-pound out-of-control airplane. As I was floating down I watched our aflame F-4 nose over and begin a steep descent. As I was watching this happen I remember how quiet and calm it was, just seconds after calamity, with the clouds floating through the air as if I was a bystander watching all this unfold.

I saw Dwyer and Kindseth, who had been in front of us, flying below, searching. They were scanning the water surface, unaware that we had climbed prior to ejection. I could also see Schumacher in his parachute below me, who fired a flare, hoping to get their attention. I remember thinking rather nonsensically to myself "Don't shoot Bill and Doug down with a flare, Frank. I don't think the Navy would be too pleased!"

During Ritchie's descent, he also saw several fishing boats leaving the island, heading in their direction. Upon hitting the water he immediately detached his parachute and climbed into the life raft. With the fishing boats still very much on his mind he began furiously paddling away from the direction he believed they would be coming. Though it seemed like much longer, the two men were in the water less than a half-hour before being rescued by Navy helicopter and taken aboard a destroyer. When they arrived back on *Enterprise* the following morning Skipper Schwartz welcomed them aboard, then promptly informed them they were on the afternoon schedule.

By the middle of April, missions were being flown into the heart of North Vietnam daily, to areas and targets previously restricted. Sorties were flown around the clock as air wings on Yankee Station rotated between noon-midnight (blue) or midnight-noon (red) schedules. The pace became so intense that the missions began blending into each other, with little discernible difference. Objectives in the heavily defended cities of Hanoi and Haiphong, along with targets in Vinh, Thanh Hoa, Hon Gai, and Kep were constantly attacked. With every Alpha strike flown the North Vietnamese zealously responded with AAA and SAMs.

With spring weather came the inescapable tropical heat of SEA, making "Condition Watch" dreaded duty for the *Fighting Falcons*. For this, a crew would sit in their aircraft on the flight deck for two to four hours, ready to launch within minutes should the need arise. With temperatures routinely registering over 90°F with accompanying humidity at 90% or greater, even with the canopy open the cockpit would quickly become a sauna, with temperatures exceeding 120°F. Couple the tropical sun and humidity with multiple aircraft belching searing hot jet exhaust and the result was misery. Kindseth recalls:

> By the time Condition Watch ended and we climbed down and made our way to the ready room, we were cooked. Between Condition 5 and stress I lost so much weight that cruise that by the conclusion my 5' 10" self weighed 129 pounds. I was not much more than a mustache and belt buckle!

On 10 May the Navy continued their assault on the infrastructure of North Vietnam that began the month before, launching another attack on the two Haiphong TPPs that provided electricity to the city. On that day CVW-9 attacked the Haiphong TPP East while attack aircraft from *Hancock* pounded Kien An Airfield and aircraft from *Kitty Hawk* attacked the Haiphong TPP West. Within the *Enterprise* strike group were 10 F-4Bs from VF-92 and VF-96, providing flak suppression and TARCAP. The mission was hugely successful with Commander Schwartz and RIO Lieutenant Commander Nordell receiving recommendations for DFCs.

The day 19 May began what would be a costly, and unforgettable week for the Navy. It was on this day that the *Fighting Falcons* lost their XO, Commander Rich, along with RIO Lieutenant Commander William Stark while on an Alpha strike to the Van Dien Truck Maintenance Depot and SAM storage facility six miles south of Hanoi.

In addition to Rich and Stark CVW-9 on *Enterprise* also lost pilot Lieutenant Commander Red McDaniel and B/N Lieutenant James Kelly Patterson, who were flying an A-6A Intruder of VA-35 *Black Panthers*. The strike group was forty miles south of Van Dien inbound to the target when multiple SAMs were launched. Doug Kindseth, who was flying with Bill Dwyer that day, remembers counting no less than seven missiles in the air. Rich and Stark call sign "Showtime 604" (BuNo. 152264) were hit in succession by two SAMs. In the process of ejecting Stark suffered compound fractures of the lower vertebrae, a broken arm, and a broken knee. He was imprisoned as a POW until released during *Operation Homecoming*. Commander Rich never made it out of the Phantom. His remains were repatriated in April 2000.

Ready to launch from *Enterprise*.

Several VF-96 officers recall sitting in the squadron ready room that day when pilot Commander Bob Glaves entered, somberly stating that Rich and Stark had been shot down, and that was all he knew. After Rich was lost Commander Joe Paulk became XO of the squadron.

In addition to the Phantom and Intruder from *Enterprise* four additional Navy aircraft were lost that day. An F-4B of VA-114 and RA-5C of RVAH-13 were lost from *Kitty Hawk* while *Bonnie Dick* lost two F-8s from VF-211 and VF-24, respectively.

The day was one of the darkest of the conflict for Naval Aviation and henceforth became known as Black Friday. Sadly, the men of the air wing had little time to absorb nor mourn the loss of their shipmates and friends as the mission to Van Dien was repeated on 21 May. On that day pilot Lieutenant Dennis Wisely and RIO Ensign Jim Laing were flying an F-4B call sign "Linfield 213" (BuNo. 153040) of VF-114 from *Kitty Hawk* in a TARCAP role. While evading multiple SAMs, they were hit by automatic weapons fire. Heading west to Thailand, both were forced to eject shortly after crossing into Laos. A Navy Sea King helicopter (BuNo. 151530) with four crewmen from *Hornet* was dispatched to rescue them, however, it too was hit by AAA over Laos, forcing the helicopter down. Fortunately and rather miraculously all six men were subsequently rescued.

The onerous and difficult task of writing VF-96 casualty forms for Navy records went to Doug Kindseth, who was the communications officer of the squadron that cruise. After garnering signatures from the CAG administrative officer and Lefty Schwartz, he would take the forms to Skipper Holloway for his signature. He states that before signing off on any casualty report Captain Holloway would spend time asking pointed questions about the man, showing a genuine interest.

Lieutenant (junior grade) Doug Kindseth

Captain James Lemuel Holloway III,[50] also known as "Jimmy Triple Sticks," "Jimmy 3" and "a little round mound of sound," was a highly respected CO, known for his care and concern of those under his command. Unlike skippers who were known for staying close to their quarters while letting the XO run the boat, Holloway was hands-on in his approach. It was not unusual for him to pop into a ready room unannounced and talk candidly with airmen, or watch a movie with them. It was also not unheard of for Holloway to appear on the flight deck as aircraft were preparing to launch, giving words of encouragement to airmen. He also enjoyed leisurely strolling down the many passageways of *Enterprise*, keeping apprised of the small city and all the lives onboard for which he was responsible.

Captain James Holloway III          Commander Richard Rich

Doug recalls one day shortly after arriving on the carrier he was traversing a passageway on the boat when Captain Holloway appeared. Never having formally met the man and following military protocol, Doug glued himself to the bulkhead in a rigid state of attention. When the skipper passed by he cheerfully remarked "Looking good today, Doug!" As it turns out Holloway studiously reviewed the personnel file of every man aboard, often remembering and addressing them by name.

Attending a Tailhook Convention in the early 2000s in Reno, Doug witnessed first-hand the respect the men had for Holloway. Hundreds of *Fighting Falcons* veterans and guests had gathered in a hotel bar for a squadron reunion. Enjoying their revived camaraderie and reminiscing of old times, the discussions soon became lively and roisterous. Admiral Holloway entered the room which immediately fell silent, with many who had been sitting rising to their feet. As he made his rounds through the crowd he recognized numerous faces. It was no surprise that he addressed many of them by name, several decades after serving together.

Terry Born remembers the last two Alpha strikes he flew in 1967, to the Paul Doumer Bridge between Hanoi and Haiphong. The first mission was a massive Alpha strike with VF-96 in a flak suppression role. As had become customary for this particular bridge, post-BDA photographs showed the structure still standing, with minimal damage. A second strike was scheduled for the following day, with Born and RIO Liscum not taking part. When the option was given for any man to bow out, due to both the danger and being the last weeks of the deployment, a married pilot quietly recused himself. Born stepped up and volunteered to take his place, much to the consternation of Liscum.

This incident was not the first time that Born had volunteered himself and Liscum to go on a mission in somebody else's place. He recalls an instance when they were a standby aircraft for two squadron Phantoms preparing to launch on a night recce mission. One of the pilots remarked over the radio that a gauge for his number two engine had a light out. Terry chimed in "I wouldn't go if I were you" which resulted in the pilot cautiously downing his aircraft. Born and Liscum went in their place, eliciting the pained question from his RIO, who had likely been hoping for an uneventful night, "Now *why* did you do that?!"

The last sorties of the deployment were flown on 20 June. There was no playing it safe as missions that day were flown into the heart of North Vietnam, from Thanh Hoa to Hanoi. After the last of the line operations had been completed and all aircraft safely recovered many parties were kicked off below deck, with cocktails abounding as the men celebrated the end of another combat cruise. Following a four-day stop in Subic Bay *Enterprise* set an easterly course to NAS Alameda, arriving on 6 July.

It would be less than six months before *Enterprise* and CVW-9 would again deploy to SEA. The only consolation was that with a 3 January 1968 departure date the air wing and ship's company would not be deprived of a third consecutive holiday season with their families.

VF-96 participated in all of the major strikes that deployment, including Thanh Hoa, Van Dien, Hai Duong, Kep, Hon Gai, Hanoi, and Haiphong. During the 132 days spent on the line, VF-96 flew over 3,067 combat hours while participating in more than 1,731 combat missions, dropping nearly 1,350 tons of ordnance on the enemy. The five Phantoms lost by VF-96 that cruise resulted in 2 POWs and 4 fatalities. For the air wing as a whole, they clocked 11,470 strike sorties which included 5,808 *Rolling Thunder* missions, losing 14 aircraft resulting in 4 POWs, 1 MIA, and 10 fatalities. Facing withering and relentless defenses protecting some of the most heavily defended targets in the world, the men of VF-96 who served on that pivotal cruise battled AAA and MiGs, missiles, and monsoons. They had endured painful losses and celebrated joyous triumphs. Lifelong bonds were created, bonds that remain strong to this day. The *Fighting Falcons* would be deployed to North Vietnam five additional times before the Navy left the water of Southeast Asia at the end of 1973, with their last deployment aboard *Constellation* as part of CVW-1 ending in October of that year. During those subsequent cruises, the squadron would not suffer a single additional casualty. This made the 19 May 1967 death of Commander Rich the last of the conflict for the squadron.

Statistically, the year 1967 represented the height of the Vietnam air war for the United States, with the Navy losing 185 aircraft resulting in 55 POWs and 124 fatalities. By the time *Enterprise* returned to Yankee Station in February 1968 the Tet Offensive had been launched, dramatically altering the course of the conflict and American politics.

On 31 March, following the shock of the offensive that reigned political hell upon him, a broken President Johnson announced a halt to bombing in North Vietnam above the 20th parallel. He then declared he would not be running for re-election.

Seven months later, just days away from the election pitting Richard Nixon against Hubert Humphrey, Johnson announced a halt to all bombing above the 19th parallel, effectively ending *Operation Rolling Thunder* after nearly four years. Missions over North Vietnam, Laos, and Cambodia would continue with *Operation Commando Hunt*, mainly an interdiction effort focusing on the Ho Chi Minh Trail. The air war over North Vietnam would not resume in full force until April 1972, when President Nixon unleashed *Operation Linebacker* in response to North Vietnam's Easter offensive that began on 30 March of that year. This was followed by *Operation Linebacker II* in December.

The Peace Accords were signed in Paris on 27 January 1973, ending the air war over North Vietnam once and for all. By that time VF-96 had gained the distinction of achieving more confirmed MiG kills than any other Navy combat squadron, a total of ten. The majority of these came on 10 May 1972 during *Linebacker* operations while flying from *Constellation*. On that prodigious day the *Fighting Falcons* downed six VPAF MiG-17 fighters. Leading the pack was pilot Lieutenant Randy "Duke" Cunningham and RIO Lieutenant (junior grade) Willy Driscoll, who downed three MiGs, bringing their total air victories to five. Also on 10 May pilot Lieutenant Matthew Connelly and RIO Lieutenant Thomas Blonsky bagged two, while pilot Lieutenant Steven Shoemaker and RIO Lieutenant (junior grade) Keith Crenshaw brought down one. The third and final victory for Cunningham and Driscoll that day resulted from a dogfight with a MiG-17 that is legendary amongst fighter pilots. Shortly after achieving this final victory their Phantom (BuNo. 155800) was hit by a SAM near Nam Dinh. Cunningham was able to nurse the crippled jet five miles off the coast before it became uncontrollable. Both men ejected and were rescued by HH-3A Sea King helicopters from HC-7.

While that May day resulted in VF-96 emerging as the Navy squadron with the most confirmed MiG kills, it also resulted in Cunningham and Driscoll becoming the only Navy aces of the conflict. For the USAF the 555th TFS (Triple Nickel)/8th TFW, flying the F-4, led in confirmed MiG kills with a total of 36. The 555th was led for a time by legendary WWII fighter ace Brigadier General Robin Olds, mastermind of the highly successful *Operation Bolo*.

Unsurprisingly, the aircraft with the most aerial victories in Southeast Asia was the Phantom, accounting for 147 VPAF aircraft shot down, a crowning achievement for the double ugly fighter that almost never was. Although the USAF is credited with downing more enemy aircraft, the Navy became so adept at aerial combat in the sans cannons Phantom that by the conclusion of the conflict they had a greater enemy aircraft kill ratio than the Air Force, 4.7:1 vs. 3.8:1. In fact, after the conflict ended the Air Force sought advice from the Navy and Marine Corps on effective dogfighting tactics in the Phantom.

The F-4 would continue its faithful service with the Navy until the inevitable evolution of military aerospace technology caught up with it. It was Saturday 18 October 1986 when the LSO aboard *America* routinely radioed "call the ball " to pilot Commander George Kraus on final approach for landing on the carrier. Minutes later the F-4S Phantom II touched down on the flight deck. As Commander Kraus of VF-202 *Superheats* shut down the engines of the mighty fighter that had thundered through the halls of military aviation, the turbines let out one final adieu. It was a moment that marked the end of an era.

## The Continuing Saga of Ensign Thornton

After landing in a dry rice paddy, Gary Thornton released his harness and gathered his parachute into a ball for concealment, hiding behind a berm between paddies. He heard yelling, and spotted a North Vietnamese man about 75 yards away, pointing an AK-47 assault rifle at him. He then realized his bright, glossy white flight helmet was like a gleaming beacon in the afternoon sun, betraying his presence. Reasoning that attempting evasion at this juncture could very well result in his death, he surrendered. Thornton had been on the ground a mere 30-40 seconds before he was captured.

Once in custody, he was stripped of his possessions, which included his clothing. He was given a black pajama-type uniform, bound, and marched under guard for several miles to a holding area. He was then forced to crawl through a pig trough on his hands and knees to a crude small cell inside the sty. That afternoon and evening he was forced out of his cell several times by what was likely the village commissar, who had a bullhorn he used to incite the baying crowd wanting to see the downed American air pirate. At about 9:00 PM that night a truck came and he was loaded into the back along with two North Vietnamese guards, beginning the trek to Hanoi. Somewhere around 3:00 AM the next morning the truck stopped at a remote mountain hamlet, where Thornton would spend the next day until continuing the journey.

Despite the dire seriousness of the situation, and with his ultimate fate still unknown, Thornton found a dash of humor that lifted his sagging spirits. When he had initially been loaded onto the truck local authorities transferred his possessions, along with a wing pylon from the aircraft, to the guards taking him to Hanoi. After stopping in the hamlet the guards began tinkering with the pan from his ejection seat. They spotted a small lanyard with a knob on the end. Curious, they pulled it and the life raft attached to the seat pan instantly and explosively inflated, shocking the guards and sending them fleeing over the side of the truck. Slowly and cautiously they peered back in, trying to ascertain exactly what happened. By this time Thornton was trying his hardest not to laugh, lest the guards take out their embarrassment and irritation on him. However, the folly was not quite complete. With the back of the truck now fully occupied by an amused American POW and an inflated life raft, one of the guards pulled out a knife to cut and deflate it. As he stabbed at the raft his dull blade failed to puncture the resilient material, resulting in the knife harmlessly bouncing away. Unable to conceal his laughter at the comedic scene any longer, Thornton demonstrated to the guards how to deflate the raft.

After resting for a few hours he was subjected to a mock trial in Vietnamese. Due to the language barrier, he is unsure of the proceedings but has little doubt that he was found guilty of whatever charges he was being tried on. That evening he was loaded on another truck and taken to the notorious Hanoi Hilton, arriving early the next morning, 22 February. Blindfolded, he was led to a cell block of the prison known as "Heartbreak Hotel". Thornton remembers that conditions at the prison were horrendous; a portrait of rampant filth, starvation, and suffering. After several hours alone in a cell, he was retrieved by his captors to begin interrogation.

American airmen had a good idea of what was occurring inside North Vietnamese POW camps, which is now well-documented. They had been briefed and trained on resisting the torturous interrogation techniques and psychological warfare meant to break their spirits, make them talk, or sign propaganda statements. The men held out as long as they were physically and mentally able. "We were bent," Thornton remembers, "but not broken."

He spent his first six or seven weeks, during his initial interrogations, in solitary confinement. Thornton would revisit solitary during his time as a POW, for infractions such as communication with other prisoners. Around 10 March he was moved from Heartbreak Hotel to the "Little Las Vegas" section of the prison, into cell block "Riviera". In the middle of April, he was moved again, to the "Thunderbird" cell block, joining Tom Storey.[51] In mid-May Jay Jayroe[52] and Mike Cronin[53] arrived, joining Thornton and Storey. On 20 May Cronin was moved out to another cell he would share with Ralph Gaither[54] and Tom Collins[55] moved in.

During the night of 6 June Thornton, Collins, and Jayroe were removed from their cell in Thunderbird, leaving Storey behind. After being blindfolded and their hands bound, they were loaded onto an open vehicle and transported across Hanoi to a new POW facility which became infamously known as "Plantation". As the men were slowly driven through the capital they were preceded by a political operative with a bullhorn, inviting locals to demonstrate their unhappiness by beating the prisoners. Thornton describes the harrowing ordeal:

We were blindfolded so we couldn't see the hits coming. I distinctly remember being hit in the back of my head so hard that I literally saw brightly colored stars and flashes! As I remember, the ride was only a couple of miles, but took about 30 minutes as they were purposefully driving very slow.

The men arrived at Plantation later that night, becoming part of the first dozen or so prisoners, making them plank owners. Any hopes for an improvement in conditions were short-lived, with their new accommodations as filthy and inhumane as Hanoi Hilton. Despite the abhorrent conditions, the men were required to sweep and clean their rooms daily. Plantation was to become a showplace for North Vietnamese propaganda such as the East German film "Pilots In Pajamas". The photo of Paul Galanti[56] showing the Navy POW alone in a cell that made the 20 October 1967 cover of LIFE magazine, was taken at Plantation. This iconic and famous photo further galvanized the American public in opposition to the conflict.

On 7 July 1967, the men were moved from Plantation to the aptly named "Power Plant" which was about 200 yards from Hanoi's main TPP, a common target of American attacks. After five to six weeks, around the third week of August, they were moved to "Dirty Bird West" on the opposite side of the power plant except for much closer, perhaps fifty yards. Twice a day a pair of prisoners would trek to the power plant to collect condensed steam for drinking water. If they were engaged in this task during one of the many air raids they were forced into vertical 36" concrete-lined pipes and covered with a lid until the raid was over.

Power Plant and Dirty Bird West facilities were not prisons; they were normal city structures. This was so the men were seen and recognized as POWs, not only to the people of Hanoi, but to the world. The reason for this was apparent—they were being used as human shields, and the North Vietnamese wanted the United States to know it. There were numerous air raids during those months on Power Plant and Dirty Bird West, with the presence of POWs not deterring the bombing effort. Gary notes:

> It was pretty scary during those bombing attacks. Our bed boards were split bamboo frames that lay on the floor. We'd hold them over ourselves to fend off plaster falling from the ceilings. There was a flak site that I never saw, but heard and felt it whenever it fired as it could not have been more than 15 yards from our cell. Those guns were very, very loud!

Thornton remembers the day following a particularly intense air raid when he and Jayroe went to the power plant to collect drinking water:

> We "chogey-poled" our five-gallon bucket to the power plant. It had an arched entry on our side, with a guard station/checkpoint just inside. When we arrived the checkpoint was rubble! A very large crater 7-8 feet deep and 15+ feet across was now the entrance to the power plant. A living anachronism was clearing the area; an ox-drawn cart with large truck tires for wheels! Great for our morale, though we thought it best to subdue our enthusiasm.

On the night of 12 October George McKnight[57] and George Coker[58] went "over the wall" and escaped Dirty Bird, greatly upsetting the North Vietnamese. They managed to travel some 15 miles before being recaptured the following day. Both men were severely tortured as punishment. As a result of their escape POWs in Dirty Bird were forced to wear leg irons at night for a week following the escape.

On 25 October, after more than four months at Dirty Bird, Thornton, Collins, and Jayroe were moved back to Little Las Vegas, to the "Stardust" cell block, joining John Heilig,[59] where they would stay for the next thirteen months. In late November 1968, with Hanoi Hilton at overflow capacity, Thornton and Jayroe were transferred to Son Tay Prison, 23 miles west of Hanoi. When they arrived they were put in a cell with six other POWs—John Frederick,[60] Wes Schierman,[61] Orson Swindle,[62] Ralph Gaither, and Dave Gray.[63] Mike Cronin, who had been with them in Hanoi Hilton, was also present.

The North Vietnamese continued their harsh and inhumane treatment of the prisoners at Son Tay. In a twisted example of irony, the men were tortured until they agreed to sign a statement declaring they were being treated humanely by their captors. One barbaric manner of coercion was to bind a prisoner to a small wooden stool with metal wire in a physically uncomfortable position. The prisoner would then be tortured by being beaten, kept awake, and denied sleep until they agreed to sign the false statement. Thornton recalls the record went to Swindle, who was tied to the stool for more than a month. He never broke.

Within a few days of the 2 September 1969 death of Ho Chi Minh treatment of American POWs began to improve. Thornton remembers that men who were being interrogated and tortured were led back to their cells, with no further abuse. Gradually prisoners received new blankets, shutters were removed from windows, allowing sunlight into their cells, and they were allowed to see one another. In the most significant reform, in Christmas of 1969, the prisoners were allowed to write letters home. The forms provided only allowed for six lines, but as Thornton recalls "Six lines was more than enough". This is especially true in his case, where he had been declared KIA on the day of his loss, and the North Vietnamese failing to acknowledge his captivity. As far as the Navy, his family, and friends were concerned, Gary had been dead for nearly three years.

It was a day in March 1970 when a Navy sedan pulled up to the Thornton home in Porterville, California and his parents were terrified at the site. It was the same type of car and the same type of men in uniform that three years earlier informed them that Gary had been killed in action. Gary's older brother Jack was serving in the Navy Reserves in the San Francisco Bay Area at the time, and they immediately feared they had lost another son in some sort of accident. To their relief and joy, the men instead informed them there was reason to believe Gary was alive and being held as a POW somewhere in or near Hanoi. They also informed them that a letter from Gary should be arriving any day. The letter arrived two days later, much to the delight of his family.

Gary was unaware that the Navy had declared him KIA until late spring/early summer 1970. A cross-compound line of communication had been established at Son Tay, allowing for the flow of information between cell blocks. From one of these adjacent blocks came a message from VA-113 A-4 pilot Lieutenant Bob Naughton, who had been shot down on 18 May 1967. The message stated that he had attended a memorial service for Gary and Russ Goodman on *Enterprise* shortly after their Phantom was shot out of the sky.

In December 1969 Mike Cronin was moved out of Son Tay, and Chuck Boyd[64] came in. A little over six months later, on 14 July 1970, the eight men, along with forty or so additional American POWs at Son Tay, were moved to Don Hoi Barracks, known to the POWs as "Camp Faith," where they were joined by Larry Carrigan.[65] When compared to Hanoi Hilton or Son Tay, the Don Hoi accommodations were luxurious. The men were housed in communal cells and there was more time outdoors. At Don Hoi they also had running water, allowing for the presence of a crude but welcome (cold) shower.

In the early morning hours of 21 November 1970 a total of 56 Green Berets and 28 aircraft manned by 92 airmen conducted a raid on Son Tay to rescue American POWs held there. Unfortunately, all American POWs had already been transferred to other camps. In the aftermath of the failed rescue attempt most if not all American POWs were moved back to Hanoi Hilton, where attempting another such rescue operation would be all but impossible. Thornton and the rest of the POWs at Don Hoi arrived back to Hanoi Hilton on 24 November. In a major shift since their internments all of the men from Don Hoi, numbering some 56, shared a large communal cell known as "Room 3" in a section of Hanoi Hilton dubbed "Camp Unity". One of Gary's cellmates in Room 3 was Navy pilot Lieutenant John "JB" McKamey,[66] who used a silent hand-signal method[67] to communicate with fellow POW Lieutenant (junior grade) Lewis "Scurvy Irv" Williams,[68] who was in "Room 2".

Williams relates that the two men stood on concrete pedestals with rolled blankets underneath their feet. This barely gave them enough clearance to see each other through the raised, barred window between the two cells. During their communication sessions, which could last hours, lookouts were posted to warn when NVN guards were coming their way. If they were caught, punishment could be severe.

His B/N Lieutenant (junior grade) Michael Christian was also in Room 2 during that time. He arrived to VA-85 in February 1967, six weeks before being shot down. Christian had graduated from high school in Huntsville, Alabama at age 16, enlisted in the Navy at age 17, and made petty officer before attending OCS and eventually A-6 B/N training. Williams describes him as a "brilliant man" who held a burning resentment and hatred of his captors. He would go out of his way to let them know how he felt, from gazes of seething anger to numerous acts of defiance. This led to multiple punishments, each of which only worked to reinforce his hostility.

During internment Christian constructed a makeshift American flag using a bamboo needle and other scavenged items. He hid it amongst his few possessions, revealing it when necessary to boost the spirits and morale of his fellow POWs. On or about Christmas Day 1972 the North Vietnamese conducted a thorough inspection of the prisoner's meager belongings and discovered the flag. This resulted in beatings and torture that almost cost Christian his life. After several weeks of being nursed back to life by fellow POWs, he defiantly began the construction of another flag.

Following his release during *Operation Homecoming* Lieutenant Commander Christian returned to active duty with the Navy. He retired in 1978 in protest over President Carter granting pardons to Vietnam draft dodgers. Tragically, he lost his life at 42 years of age in a house fire in Virginia Beach on Labor Day weekend, Sunday 4 September 1983.

*Operation Homecoming* commenced in February of 1973, shortly after the Peace Accords were signed in Paris. Sick and wounded were released first, followed by those who had been held the longest. Thornton walked out of Hanoi Hilton on 4 March, along with Swindle, Williams, Christian, and others in that release. They were then transported to Gia Lam Airport in Hanoi where they boarded an Air Force Lockheed C-141A Starlifter christened the "Hanoi Taxi," flying to Clark Field in the Philippines, and freedom. Of his time in captivity Thornton remarks:

> We all knew there was a possibility of being killed or captured. That was part of the business we were in and I accepted that. The most distressing thing to me about being a POW was the length of time. I am not bitter about being a prisoner but disappointed that I spent six years in captivity. Those are years of my life I will never get back.

After a rehabilitation period, Thornton chose to stay in the Navy, ending up in a Lockheed P-3 Orion patrol squadron out of Moffett Field in the San Francisco Bay Area. He retired from the Navy as a commander after 32 years of service; 28 in active duty and 4 in reserves. Thornton was the last naval aviator of the conflict to serve six full years as a POW.

---

[38] Lieutenant (junior grade) Charles Clydesdale was flying an A-1H Skyraider (BuNo. 135375) of VA-95 *Green Lizards* from *Ranger.* As he was returning to the carrier following a mission the engine in his aircraft lost power. He was killed while attempting to ditch his Skyraider in the gulf.

[39] 1st Lieutenant George Smith was flying a F-100D Super Sabre (Serial No. 55-3625) of the 401st TFW/613th TFS from Da Nang, and was on his second attack run when his aircraft was hit by AAA. Smith was killed when he was unable to eject from his damaged aircraft before it impacted the ground.

Lieutenant Commander Raymond Vohden was flying an A-4C Skyhawk (BuNo. 148557) call sign "Diamondback" of VA-216 *Black Diamonds* from *Hancock*. He was attacking the bridge at Dong Phuong Thong, ten miles north Dragon's Jaw when his aircraft was hit by small arms fire, causing a fuel leak. He ejected and was captured, subsequently released during *Operation Homecoming*.

[40] Captain Walter Draeger A-1H Skyraider Det 10 1131st SAS Bien Hoa—KIA. Captain Carlyle Harris F-105D Thunderchief (Serial No. 62-4217) 44th TFS/18th TFW Korat—POW. Major Frank Bennett F-105D Thunderchief (Serial No. 59-1754) 354th TFS/355th TFW Korat—KIA (VPAF MiG). Captain James Magnusson F-105D Thunderchief (Serial No. 59-1764) 354th TFS/355th TFW Korat—KIA (VPAF MiG).

[41] During the conflict the Chinese would employ signal "meaconing," an electronic warfare operation meant to disrupt and confuse aircraft navigation systems in the Tonkin Gulf. The objective was to trick American military aircraft into Chinese airspace, often that surrounding Hainan Island, where they could be openly attacked or forced down with little fear of challenge or reprisal.

[42] The first non-combat (operational) loss of an F-4 in the Vietnam theatre occurred on 13 November 1964. Lieutenant W.R. Moore and RIO Lieutenant (junior grade) W.M. Myers of VF-142 from *Constellation* were practicing air combat maneuvering when their F-4B (BuNo. 151402) stalled and entered a spin. Both men ejected and were rescued by Navy helicopter.

[43] The first credited partial kill of a MiG by American aircraft in the Vietnam conflict occurred just five days prior, during the first wave of attacks on Dragon's Jaw Bridge. Air Force Captain Donald Kilgus of the 416th TFS/3rd TFW from Da Nang, flying a F-100D Super Sabre (Serial No. 55-2894) shot down a MiG-17 with 20mm cannon fire. Kilgus served four tours to Southeast Asia, including stints as a Misty FAC and Wild Weasel.

---

[44] Though officially credited, there is some question as to whether Murphy and Fegan actually shot down a Chinese MiG-17 during the dogfight. Some sources claim F-4 pilot Lieutenant (junior grade) Dave Batson and RIO Lieutenant Rob Doremus of VF-21 *Freelancers* from *Midway* achieved the first confirmed Navy MiG kill of the conflict, shooting down a MiG-17 with an AIM-7 Sparrow on 17 June 1965.

[45] On 15 March Lieutenant (junior grade) Charles of VA-95 call sign "Fortress 512" was flying an A-1H Skyraider (BuNo. 135375) on a mission against an ammunition depot at Phu Qui. Returning to *Ranger*, the aircraft lost engine power for unknown reasons. As the Skyraider did not have an ejection seat Clydesdale ditched the aircraft in the sea and submerged with it, despite a rescue effort by a helicopter from the destroyer USS *Wiltsie* (DD-716).

On 11 April Lieutenant (junior grade) William Swanson of VA-95 call sign "Fortress" was flying a Douglas A-1H Skyraider (BuNo. 135226) from *Ranger* on a *Steel Tiger* mission over southern Laos. The flight of two Skyraiders were east of Ban Senphan when they encountered AAA fire. Swanson's aircraft took several hits and began trailing smoke while descending to the jungle below. His wingman tried to contact him on the radio however there was no response, leading to the conclusion that he was incapacitated or killed by AAA. His remains were recovered in 2009-2010.

[46] Captain Richard "Pop" Keirn was a veteran of World War II, where he had served as a B-17 co-pilot with the "Bloody 100th" bomb group of the 8th Air Force. On 11 September 1944 the B-17 he was co-piloting (Serial No. 43-38043) was shot down near Leizpig and Keirn was taken POW. Released at the conclusion of the war, he was recalled to active duty in 1956 and flew the North American F-86 Sabre and F-100 Super Sabre. He was on his eleventh WSO mission in Vietnam when he was shot down. Keirn is one of only two Vietnam veterans to have been held as a POW by both Nazi Germany and the DRV.

[47] On 18 July 1965 Navy Commander Jeremiah Denton and B/N Lieutenant (junior grade) Michael Tschudy of VA-75 *Sunday Punchers* from *Independence* call sign "Flying Ace 500" were flying an A-6A (BuNo. 151577) on a mission to the Ham Rong port facilities near Thanh Hoa. One of their Mk.82 bombs exploded prematurely, badly damaging the Intruder and forcing their ejections. Both men were taken as POW. On 2 May 1966 Denton was forced by the North Vietnamese to take part in a televised propaganda broadcast. While being filmed he slyly blinked T-O-R-T-U-R-E in morse code, letting the world know what was happening to American POWs inside NVN POW camps. Denton and Tschudy were released during *Operation Homecoming*.

---

48 The A-3 was a tanker, also known as a "Whale". Reconnaissance and electronic warfare aircraft were sometimes referred to as "Q Birds". Thus the reconnaissance version of the A-3, the RA-3B, was sometimes referred to as a "Q Whale".

49 The transition to under-seat rocket-motor ejection had not yet taken place, meaning Thornton experienced an instantaneous 24g force on his body. This MK5 model of Martin-Baker ejection seat was often referred to as "Martin-Breaker" due to the orthopedic injuries often suffered from its use.

50 Captain Holloway would later to rise to admiral and the 20th CNO (1974-1978). He passed away 26 November 2019 age 97.

51 On 16 January 1967 Air Force Captain Thomas Storey and PSO 1st Lieutenant Ronald Mastin of the 11th TRS/432nd TRW from Udorn call sign "Outlaw" were flying an RF-4C (Serial No. 65-0818) reconnaissance Phantom to obtain targeting intelligence. They were hit by AAA 25 miles north of Kep Airfield and ejected. The men were on their 34th combat mission together when they were shot down and captured. Both were released during *Operation Homecoming.*

52 On 19 January 1967 Air Force Captain Julius "Jay" Jayroe and WSO 1st Lieutenant Galand Kramer of the 390th TFS/366th TFW from Da Nang call sign "Rattler" were flying an F-4C (Serial No. 64-0845) as part of a formation of four Phantoms escorting two RF-101 Voodoo reconnaissance aircraft of the 20th TRS on a mission to Kep Airfield. The North Vietnamese responded with heavy AAA fire as well as multiple SAMs. Jayroe and Kramer were hit, ejecting from their Phantom five miles east of Bac Ninh. They were both captured, becoming POWs until released during *Operation Homecoming.* Jayroe was on his 100th and final combat mission before his scheduled return to the States.

53 On 13 January 1967 Navy Lieutenant (junior grade) Michael Cronin of VA-23 *Black Knights* from *Coral Sea* call sign "Law Case 345" was flying an A-4E Skyhawk (BuNo. 151158) on an attack against railway cars at Vin Loc, 30 miles south of Thanh Hoa. Diving on the target, he was at 4,000 feet and about to release his Mk.82 bombs when he was hit by AAA. The aircraft began burning, eventually breaking into two sections. Cronin ejected feet wet, but a strong on-shore wind blew him back to landfall. He was captured and imprisoned, released during *Operation Homecoming.*

54 On 17 October 1965 Navy Ensign Ralph Gaither and RIO Lieutenant (junior grade) Rodney Knutson of VF-84 *Jolly Rogers* from *Independence* call sign "Victory 215" were flying an F-4B Phantom (BuNo. 151494) on an Alpha strike mission against the Thai Nguyen road bridge 30 miles north of Hanoi. They were egressing from the target area when they were hit by AAA just a few miles from the Chinese border. Both men ejected, with Knutson suffering painful injures to his neck and back. After landing and being fired upon Knutson drew his .38 service revolver, killing two North Vietnamese soldiers before being shot himself and subdued. Both men were captured and released during *Operation Homecoming*. The raid on the Thai Nguyen Bridge was costly for the Navy and *Independence*, losing three Phantoms from CVW-7 (2 VF-84, 1 VF-41) resulting in four POWs and two fatalities.

55 On 18 October 1965 Air Force Captain Thomas Collins and WSO 1st Lieutenant Alan Brudno of the of the 68th TFS)/8th TFW from Korat call sign "Argon" were flying an F-4C (Serial No. 64-0730) on a mission attacking a rail bridge 35 miles south of Vinh. They had just begun a dive on the target when they were hit by AAA, resulting in their aircraft becoming uncontrollable, necessitating their ejection. Both men were released during *Operation Homecoming*. After enduring more than seven years of brutal, inhumane treatment and illness, on 3 June 1973, four months after his release from captivity, Brudno committed suicide. He was a graduate of Massachusetts Institute of Technology, and had planned to apply to NASA's astronaut program following his Air Force service. In April 2004 his name was added as a casualty of the conflict to the Vietnam Veterans Memorial in Washington, DC.

56 On 17 June 1966 Navy Lieutenant (junior grade) Paul Galanti of VA-216 from *Hancock* call sign "Black Diamond 693" was flying an A-4C Skyhawk (BuNo. 149528) on a strike against a railway yard at Qui Vinh, 30 miles south of Thanh Hoa. His aircraft was hit by small-arms fire, causing a fire which resulted in the Skyhawk become uncontrollable. He had almost made it to the coast and awaiting rescue forces when he was forced to eject feet dry. Galanti, who was on his 97th combat mission, was captured, spending nearly seven years in captivity before being released during *Operation Homecoming*.

57 On 6 November 1965 Air Force Captain George McKnight of the 602nd SOS (ACS)/6251 TFW from Udorn call sign "Sandy 14" was flying an A-1E Skyraider (Serial No. 52-132439) on a rescue mission, trying to recover Lieutenant Colonel George McCleary who had ejected the previous day while flying a F-105D Thunderchief (Serial No. 62-4342) on an Iron Hand SAM-suppression mission near Hanoi. McKnight was trying to assist in the rescue of another Skyraider that had been shot down during the recovery attempt, when he too was hit by AAA approximately seven miles north of Dragon's Jaw Bridge. The rescue attempt of McClearly, who was ultimately declared KIA, cost the Air Force three aircraft (2 A-1E Skyraiders and 1 CH-3C Jolly Green Giant), resulting in five POWs for North Vietnam, all of whom were repatriated during *Operation Homecoming*.

58 On 27 August 1966 Navy Lieutenant Commander John Fellowes and Bombardier/Navigator Lieutenant (junior grade) George Coker of VA-65 *Tigers* from *Constellation* call sign "Cupcake 405" were flying an A-6A Intruder (BuNo. 151822) on a mission against the Ngoc Son Bridge. They were 15 miles northwest of Vinh when they began a climb to commence their bomb run. A burst of AAA blew a portion of their starboard wing away, with what remained catching fire. With extensive damage to the wing and loss of controls the aircraft entered an inverted spin, with the men ejecting from the mortally damaged Intruder at 3,000 feet. Both were captured shortly after hitting the ground and released during *Operation Homecoming*.

59 On 5 May 1966 Navy Lieutenant John Heilig of VFP-63 from *Hancock* call sign "Cork Tip 908" was flying an RF-8A Crusader (BuNo. 146831) on a photoreconnaissance mission over the Song Ca River, 20 miles northwest of Vinh. The Crusader was hit by 37mm AAA, resulting in an uncontrollable roll. Heilig ejected and was captured, subsequently spending nearly seven years as a POW before being released during *Operation Homecoming*.

---

[60] On 7 December 1965 Marine Corps Major pilot John Dunn and RIO Chief Warrant Officer (CWO) John Frederick of VMFA-323 *Death Rattlers* from Da Nang were flying an F-4B Phantom (BuNo. 152261) escorting an EF-10B Skyknight on a night mission over North Vietnam when they were shot down 45 miles west of Thanh Hoa. It is suspected they were hit by a SAM. Both men ejected with Dunn, who was the XO of VMFA-323, evading the enemy for six days before being captured. Frederick was also captured, enduring nearly seven years of torture and gross mistreatment until he passed away in captivity from typhoid fever on 19 July 1972. For heroism during internment Frederick was posthumously award the Navy Cross. Dunn was released during *Operation Homecoming*.

[61] On 28 August 1965 Air Force Captain Wesley Schierman of the 67th TFS/6234th TFW from Korat call sign "Elm 1" was flying a F-105D Thunderchief (Serial No. 63-8282) on an armed reconnaissance mission when he commenced an attack against a military barracks 105 miles west of Hanoi, near the Laotian border. On his second strafing run his Thud was hit by small arms fire, forcing him to eject. Schierman, who was on his 37th combat mission, was captured and spent over seven years as a POW before his release during *Operation Homecoming*.

[62] On 11 November 1966 Marine Corps Captain Orson Swindle of VMF-235 *Death Angels* from Da Nang was flying an F-8E Crusader (BuNo. 150858) on a mission against military buildings two miles north of the DMZ. His aircraft was hit by AAA, forcing him to eject near Vinh Linh. Swindle was on his 205th and last scheduled combat mission when shot down, and was released during *Operation Homecoming*.

[63] On 23 January 1967 Air Force 1st Lieutenant Barry Bridger and WSO 1st Lieutenant David Gray of the 497th TFS/8th TFW from Ubon call sign "Shark" were flying an F-4C Phantom (Serial No. 64-0773) on a MIGCAP mission near Son Tay, 15 miles northwest of Hanoi. Their aircraft was hit by a SAM, causing it to break into pieces. Both men ejected and were captured, spending six years as POWs until released during *Operation Homecoming*. Bridger was on his 53rd mission when he was shot down. Gray, a recent addition to the squadron, was only on his second mission.

64 On 22 April 1966 Air Force Captain Charles "Chuck" Boyd of the 421st TFS/ 388th TFW from Korat call sign "Fir" was flying an F-105D Thunderchief (Serial No. 62-4409) on an attack against a SAM site near the Song Lo River, 35 miles northwest of Hanoi. His Thud was hit by AAA, forcing him to eject. He was quickly captured and spent the remainder of the conflict as a POW until released during *Operation Homecoming*. He was on his 105th combat mission when he was shot down and captured. The 22 April ejection was Boyd's second in less than a month. On 26 February he had been flying a Thud (Serial No. 61-0215) call sign "Cactus" on a mission against a target near Ban Dang in southern Laos when he was hit by ground fire. He ejected and was rescued by a USAF HH-3C Sea King helicopter.

65 On 23 August 1967 Air Force pilot Captain Larry Carrigan and WSO 1st Lieutenant Charles Lane of the 555th "Triple Nickel" TFS/8th TFW from Ubon were flying an F-4D Phantom (Serial No. 66-0247) on a strike against the Yen Vien railway yard just north of Hanoi. They were one of four 555th Phantoms call sign "Ford" accompanying nine F-105 Thunderchiefs call sign "Falcon". As they approached the target the Phantoms were attacked by VPAF MiG-21 fighters. Carrigan and Lane, as well as another F-4 (Serial No. 66-0238) were shot down by MiG Atoll missiles. A third Phantom (Serial No. 65-0726) was shot down by AAA. The fourth Phantom (Serial No. 66-0260) of Ford flight, badly damaged by AAA over the target, barely made it back to Thailand before the crew was forced to eject. In friendly territory with no risk of capture, both were picked up by a HH-3E from Udorn. After three days of evading enemy forces Carrigan was captured and imprisoned until released during *Operation Homecoming*. Subsequent intelligence indicated that Lane was also captured as a POW, however he was not among those repatriated at the conclusion of the conflict. His fate remains unknown. The final tally of losses for Triple Nickel that day were 4 POWs and 2 fatalities.

66 On 2 June 1965 Lieutenant John McKamey of VA-23 call sign "Law Case" from *Midway* was flying an A-4E Skyhawk (BuNo. 151161) on an armed reconnaissance flight northwest of Vinh. It was his 21st combat mission. While flying low over the Song Ca River searching for an arms-transporting ferry his aircraft was hit by AAA and caught fire. He ejected and was captured, spending nearly eight years as a POW.

---

[67] When not able to physically see each other POWs used a tap code to communicate. It was based upon a 25-letter alphabet arranged in five rows and five columns. The letters "C" and "K" were combined (first row, third column). For example "God Bless You" which was condensed to "GBU" would be *tap tap, tap tap* (second row, second column) followed by *tap, tap tap* (first row, second column) and *tap tap tap tap, tap tap tap tap tap* (fourth row, fifth column). Tap code could be used to communicate not only by knocking on walls and other objects but also by coughing, sneezing, grunting and other seemingly innocuous sounds. The system was developed by USAF Captain Carlyle "Smitty" Harris who was flying a F-105 Thud when he was shot down on 4 April 1965. Harris spent 2,781 days in captivity.

[68] On 24 April 1967 Lieutenant (junior grade) Lewis Williams and B/N Lieutenant (junior grade) Mike Christian of VA-85 *Black Falcons* call sign "Buckeye 512" from *Kitty Hawk* were flying an A-6A Intruder (BuNo. 152589) as part of a 27-aircraft Alpha strike on Kep Airfield. At about 1700 hours their Intruder was hit by 85mm AAA in the port wing root, five miles east of the target. With the aircraft ablaze Williams and Christian jettisoned their munitions and tried to get feet wet. The damaged aircraft had descended from 6,500 feet to 1,500 feet when their hydraulics failed and the Intruder began to tumble, forcing their ejection. After making landfall and releasing his harness Christian immediately began running and was shot in his left thigh. This injury, along with a broken left hand sustained during ejection, limited his evasion and thus he was captured first, followed by Williams.

# Chapter V

# Tales of the Whale

Among the modern and technologically advanced aircraft operating from *Enterprise* and other carriers during the Vietnam conflict was an early stalwart of Navy jet aviation—the Douglas A-3 Skywarrior.

The aircraft was designed in the immediate years following WWII as a subsonic carrier-based nuclear bomber. Much the same as the subsequent supersonic A-3J/RA-5C Vigilante, the success of the Navy's Polaris missile program negated that intended role. Fortunately, the versatile Skywarrior found new purposes, serving throughout the course of the conflict and decades beyond, leaving a memorable mark on the history of Naval Aviation.

It was in August 1948 when the Navy released a bid for a carrier-based, long-range jet bomber capable of carrying 10,000 pounds of conventional or nuclear weaponry, with an MTOW of 100,000 pounds. Among the aerospace firms bidding on the project was Douglas Aircraft, with their effort led by famed chief designer Ed Heinemann, who was also the father of the SBD Dauntless, A-20 Havoc, and A-26/B-26 Invader as well as the A-1 Skyraider and A-4 Skyhawk. Among his many legendary creations, he considered the A-3 his finest work.[69]

Heinemann wisely opted for a lighter airframe than outlined in the Navy's requirements. He envisioned an aircraft that could operate from existing WWII-era carriers such as *Hancock* and *Intrepid*, not just the United States-class "Supercarriers" that were being planned and subsequently canceled. Achieving an MTOW of 78,000 pounds, the Douglas design was chosen in July 1949, with the company awarded a contract for two airworthy prototypes, as well as one static airframe.

The original crew configuration of the Skywarrior was three men— pilot and NFO B/N in the front seats and an enlisted petty officer seated directly behind the pilot. Though the NFO was not a pilot and did not have flight controls, he was nevertheless a de facto copilot with the same responsibilities such as reading checklists, managing fuel and tanker operations, navigation, radio communications, and ECM. As with the A-6 Intruder, the right-hand B/N seat in the A-3 was lower and behind the pilot to not impede his visibility out of the starboard side. The third-seat petty officer faced aft and acted primarily as a gunner, remotely operating a twin 20mm turret located in the tail. This petty officer also acted as the plane captain who would supervise fueling and was also qualified to perform minor maintenance on the aircraft.

To save weight and simplify cockpit design there were no ejection seats on the Skywarrior, leaving it as the Navy's only jet-powered carrier aircraft operating without this lifesaving feature. Instead, there were two parallel belly hatches, an upper and lower, on the deck of the cockpit, next to the third crewman. This was also the method used for routine cockpit crew entry and exit. In addition, there was an overhead canopy hatch in the cockpit that was routinely left open during takeoff and landing should an emergency arise. In contrast, the Air Force version of the Skywarrior, the B-66 Destroyer, of which 294 were manufactured, featured ejection seats for all crew members.

While ejecting out of an aircraft took a fraction of a second, egressing out of a Skywarrior took longer. Unlike ejection, egressing was not an instantaneous process as the crewmen had to unbuckle and unstrap numerous harnesses and belts before being able to leave their seats, a time-consuming task. To egress from the overhead hatch the crew would activate compressed gas cylinders to blow it open. For emergency egress from the belly hatches four compressed gas cylinders would force them open, forming a continuous metal slide that extended at a 60° angle from the belly of the fuselage. The purpose of the slide was to ensure that the crew cleared the fuselage, and did not get tumbled by the slipstream. Using the slide the crewmen would exit facing aft, with their parachutes preset to automatically deploy at 14,000 feet.

The canopy hatch was not recommended for egress in-flight due to the danger of striking the stabilizers or port-side fuel vent mast. There was an incident involving NFO Lieutenant Floyd Stokes, who on 8 April 1968 egressed from the canopy hatch during carrier qualifications off the coast of Southern California, the result of an in-flight emergency that was eventually brought under control. After the aircraft landed scuff marks from his boots were found on the vertical stabilizer, leading the other crewmen to believe he had been fatally injured. Fortunately, Stokes survived unscathed and was rescued from the water by a plane guard helicopter from *Constellation*. After the incident he turned in his wings, choosing to stay on the ground as an intelligence specialist.

The NATOPS manual recommended a minimum altitude of 8,000 feet above ground level and airspeed of no more than 250 knots for safe bailout from the belly hatches. If bailing out below 1,000 feet above ground survival was questionable due to the time it took for a parachute to fully deploy. In contrast, the envelope for ejection seats was much more generous.

Despite repeated calls to equip the Skywarrior with "bang" seats the original design remained throughout the life of the aircraft. This lack of ejection seats was to plague the A-3 throughout its service life, costing lives and leaving a black mark on an otherwise remarkable and highly versatile aircraft. The lack of ejection seats and several fatal incidents led to the Skywarrior being morbidly referred to as "All 3 Dead" or "Killed All 3 Dead" based upon pre-1963 A-3D/KA-3D designations.

A prototype designated XA3D-1 (BuNo. 125412) made its maiden flight in the fall of 1952, with the first A3D-1 flight occurring in February 1953. The aircraft entered fleet service in the spring of 1956 with heavy attack squadron VAH-1 *Smokin' Tigers* at NAS Jacksonville. The following month VAH-2 *Royal Rampants* at NAS North Island in San Diego became the first PACFLT squadron to receive the aircraft.

The first carrier deployment for the A3D-1 (A-3A) was in November 1956 aboard *Forrestal* during the Suez Crisis. Several months later, in January 1957, the carrier took a compliment of A-3s on its Mediterranean cruise. This second deployment largely served as a test bed for the aircraft, surveying the feasibility of routine carrier operations. Soon after, upgraded A3D-2 (A-3B) models began reaching the fleet. The refinements included more powerful Pratt & Whitney J57-P-10 turbojets, an improved bomb bay, an aerial refueling probe, and provisions for a tanker package. The last 21 of the A-3B models manufactured also saw the tail turret removed to accommodate EW equipment, as well as the installation of an upgraded Norden ASB-7 radar that featured a 300-mile range. Modifications were also made for a fourth crewman. The production run of 282 aircraft ended in January 1961, nearly four years before the type would see combat in the skies over Southeast Asia.

At more than 76 feet long, a 72-foot wingspan, and 78,000 pound MTOW, the A-3B was at the time the largest, and heaviest aircraft routinely operating from an aircraft carrier. Much of the girth was necessary to accommodate the "Fat Man" nuclear bomb, which in a time before warhead miniaturization was exceedingly large and heavy. This required the bomb bay to be no less than 183 inches in length and 66 inches in width. With its immense size and portly fuselage the Skywarrior was bestowed the apt nickname "Whale".

It was not long after the aircraft entered fleet service that the Navy smartly began diversifying its role, as ICBM technology was fast becoming the primary and preferred means of nuclear munition delivery to Cold War adversaries. Six variants of the Skywarrior served in the Vietnam conflict—A-3B bomber, KA-3B tanker, EA-3B ELINT, RA-3B reconnaissance, and EKA-3B ECM/tanker. There was also a TA-3B training/VIP transport variant which had six seats in the former bomb bay, some featuring desks. Despite all A-3 variants being derived from the same airframe, there were significant differences, especially in pressurization. In the A-3B, KA-3B, and EKA-3B only the cockpit was pressurized. In these models, if a crewmen needed to enter the bomb bay area when flying above a certain altitude the cockpit had to be depressurized first. EA-3B, RA-3B, and TA-3B models were pressurized to the rear bulkhead, which included the bomb bay area.

The ELINT A3D-2Q (EA-3B) was a PECM aircraft with the electronics package located in a canoe fairing on the belly of the fuselage, covering the area that was originally bomb bay doors. Antennas for PECM gear were located in the canoe fairing as well as in the horizontal stabilizers. The purpose of this equipment was to passively detect, classify and locate enemy radar.

An EA-3B could accommodate a crew of up to seven men. Within the cockpit were the pilot and B/N in the left and right seats, with the third and fourth seats often occupied by intelligence "spooks". These men were also Navy, TDY from intelligence units, with their presence and capabilities depending upon the mission and country the aircraft was working. Located in what was formerly the bomb bay were two to three enlisted signal evaluators and their NFO supervisor. These men sat in front of consoles that were a plethora of switches, dials, knobs, lights, and scopes. To accommodate the increased crew the EA-3B added an upper escape hatch in the evaluator's compartment, as well as a starboard side hatch for emergency egress. All those flying in the EA-3B were required to have "Special Intelligence" security clearance and mission briefings were held in a secure compartment.

EA-3Bs flew over the Tonkin Gulf and the coastline of North Vietnam as well as over Laos and Cambodia, gathering intelligence on NVN air defenses. A common mission involved flying along the coastline of North Vietnam at an altitude of 25,000-31,000 feet and some 60 miles offshore, listening for Spoon Rest and Fan Song radars of the SA-2. When these radar frequencies were detected the types and locations were recorded, with their findings broadcast over the guard channel to other aircraft. If emanating from the Hanoi area the location would be described as "Bullseye". Though tasked with many dangerous missions, not a single EA-3B airframe was lost in the Vietnam theatre.

The A3D-2P (RA-3B), a photoreconnaissance variant, could carry up to twelve vertical and oblique cameras. In addition to the pilot and NFO, there was a third crewman, a rated enlisted petty officer who bore responsibility for the photographic equipment. Among his duties were ensuring proper operation of the reconnaissance cameras and changing film magazines. He was also responsible for manually opening and closing mechanical viewing windows for the numerous cameras. These viewing windows made the RA-3B readily distinguishable from other Skywarrior models.

The Vietnam conflict was not the first time the Skywarrior was flown into harm's way as the RA-3B had flown reconnaissance missions over Cuba during the missile crisis in 1962. Among the naval aviators who flew those missions was future astronaut Lieutenant Roger Chaffee,[70] who made the flights as part of VAP-62 *Tigers* based at NAS Jacksonville. In addition to Chaffee, future astronauts Edgar Mitchell flew the Skywarrior as part of VAH-2, while Ken Mattingly flew the aircraft as part of VAH-11 *Checkertails*.

While many early RA-3B missions were cartographic flights over South Vietnam at altitudes that kept them out of range of AAA, they would later begin flying mapping and reconnaissance missions over North Vietnam, Cambodia, and Laos using still and video cameras. Nighttime recce missions utilized IR cameras to spot heat signatures of vehicles on the Ho Chi Minh Trail for attack and fighter aircraft. For these missions several RA-3Bs were painted in camouflage liveries.

In addition to modifications resulting in the EA-3B and RA-3B, in 1967 a total of 85 A-3Bs were retrofitted as KA-3B tankers, which had a range of 2,000 miles and could carry 35,000 pounds (5,147 gallons) of JP-5[71] jet fuel. Although the majority of these conversions were done at Navy rework facilities such as NAS Alameda and NAS Cubi Point, the tanker package could be installed while aboard a carrier underway. In what would make it one of the most multifaceted aircraft of the Vietnam era, 39 of these tankers were further modified to EKA-3B models. This allowed the Skywarrior to act as both tanker and ECM aircraft within the same mission.

The EKA-3B held a crew of three and incorporated two active ECM systems. The ALQ-92 A-Band communications jammer used numerous external radomes, the most prominent being four steerable fuselage side "blisters," two on each side fore and aft, and were operated by the B/N in the right seat. This system gave the EKA-3B the capability to disrupt radio frequencies, including those used by the SA-2 SAM and VPAF MiG interceptors. The use of these radio communication jammers was sometimes restricted when intelligence needed to monitor unfettered frequencies for intelligence-gathering purposes.

The ALT-27 multi-band automatic jamming and deception system sensors were located in the canoe fairing on the undercarriage of the fuselage and were operated by an NFO in the third seat. This system gave the EKA-3B the capability to degrade and "fog" enemy radar used by the SA-2 SAM. The ALT-27 not only made it difficult for SAM operators to acquire targets, but it also blinded them to incoming munitions such as HARM missiles. Only two crewmen, a pilot and NFO, were needed to man the EKA-3B if the mission did not involve ECM. However, B/Ns routinely monitored the ALQ-92 radio frequency jammer on all missions, including two-man tanker sorties.

Captain Ed Gibson flew as an NFO with VAQ-135 *Black Ravens* Det. 1 attached to CVW-11 on *Kitty Hawk*. He remembers flying a tanker mission off the coast of North Vietnam on 6 May 1972 when he detected radio transmissions from a flight of four VPAF MiG-21 fighters. He used the ALQ-92 on the EKA-3B (BuNo. 142650) to jam their radios, while two F-4J Phantoms of VF-114 from *Kitty Hawk* were vectored to intercept by an E-2 Hawkeye. In an engagement lasting only 90 seconds, former Top Gun instructor Lieutenant Commander Kenneth "Pete" Pettigrew and RIO Lieutenant (junior grade) Michael McCabe shot down one of the MiGs while his wingman Lieutenant

306

Robert Hughes and RIO Lieutenant (junior grade) Adolph Cruz bagged a second. With two of their compatriots shot out of the sky the remaining MiGs scampered away. Both Phantoms accomplished their aerial victories with AIM-9 missiles. As Ed remembers it:

> I had the ALQ-92 receiver/jammer, which was directly in front of me on the radar console, turned on during every flight. I believe there was an audio signal as well as a visual spike on the display, which caused me to immediately turn on the jammer function. The spike would pop up whenever the NVN controllers would broadcast, but the jamming was continuous until they went quiet. I think we heard that the MiGs had been shot down before we trapped, but it wasn't until after we recovered aboard and Pettigrew and Hughes landed safely on *Kitty Hawk* that congratulations were offered. Pettigrew acknowledged that our jamming helped make the aerial victories possible. That was tremendously satisfying for me.

Hughes was the son of Rear Admiral William Hughes, USN (ret.), and a 1967 graduate of USNA. Tragically, he was killed on 30 January 1979 in a mid-air collision during training off the coast of Southern California.

Skywarrior squadrons, typically consisting of 15 aircraft, were not often deployed in standard Navy fixed-wing fashion. They could be divided into numerous detachments, each attached to a different air wing. A detachment commonly consisted of three aircraft, five pilots, and seven NFOs along with approximately seventy support personnel in administration, maintenance, and operations. Each detachment had a designated OinC who, in the absence of the squadron CO, acted as skipper. An OinC customarily held a rank of lieutenant commander or commander.

EKA-3B of VAQ-135 Det 3 from *Coral Sea*.

A-3B, KA-3B, and EKA-3B detachments not attached to newer and larger supercarriers such as *Enterprise, Kitty Hawk,* and *Ranger* were not billeted onboard the boats. They instead billeted and operated from land bases such as FASU Da Nang in South Vietnam, NAS Cubi Point in the Philippines, and NAS Agana in Guam. This was necessary due to the size and weight of the Skywarrior, which even though featured folding wings and vertical stabilizer, one A-3 had nearly the equivalent physical footprint of two A-4s, A-7s, or F-4s. Slight in size when compared to supercarriers, modified WWII-era *Essex*[72]-class carriers such as *Bon Homme Richard, Oriskany, Hancock,* and *Ticonderoga* had land-based A-3 squadrons. EA-3B and RA-3B detachments were not normally billeted aboard carriers. Nevertheless, they routinely trapped aboard for refueling, pilot currency, and other purposes.

Skywarrior pilots accustomed to supercarriers who trapped aboard these *Essex*-class carriers remarked it felt like somebody had sawed the flight deck in half. In fact, if an A-3 hooked the number four arresting wire when landing on an *Essex*-modified carrier, by the time the Whale came to a stop the tip of the refueling probe protruded over the edge of the flight deck.

A fortunate by-product of the Skywarrior's size was the ability to haul cargo (both official and unofficial) and personnel, adding freighter and shuttle to the aircraft's many missions. KA-3 pilot Commander Howard Nickerson recalls a prime example of this capability which occurred in early January 1972, following the conclusions of the Bangladesh Liberation War and Indo-Pakistani War. *Enterprise* had been ordered to the Bay of Bengal in a show of force during a time when the Navy's mainstay of COD operations, the C-2 Greyhound, had been grounded due to a recent spat of fatal incidents. Beginning on 2 January 1972 a total of three KA-3s from VAQ-208 *Jockeys* made daily logistics flights from Cubi Point, with a refueling stop at U-Tapao Royal Thai Navy Airfield, where the flight crews were also refreshed. These critical logistics flights continued through 10 January, keeping the carrier resupplied during the tense situation.

Skywarriors also acted as navigational pathfinders, escorting Navy and Marine Corps fighter and attack aircraft across the vast Pacific Ocean to and from Cubi Point. The Skywarrior was chosen for this task due to the large amount of fuel it could carry as well as the ability to refuel other aircraft. Due to the vast distances involved the aircraft would make refueling stops along the way in Hawaii, Midway, Wake, and Guam. Kwajalein and Johnson Atolls were also sometimes used as refueling and rest stops during the long, trans-oceanic journeys. These flights took place during daylight hours, with the crews staying overnight at the remote outposts before continuing their journey the next day, weather permitting. There were occasions when mechanical issues precluded an aircraft from continuing its transit, leaving the crews stranded at these isolated locations for days and in some cases weeks until the issues were resolved.

A number of A-3 missions early in the conflict 1965-1967 included that of conventional bomber. It was initially utilized in this role against targets in both north and south, as well as maritime mining. The first of these bombing missions occurred on 29 March 1965 by VAH-2 from *Coral Sea*, dropping iron ordnance on Bach Long Vi Island. The last of these missions took place the first week of March 1967 with the deployment of maritime mines in NVN rivers. In these offensive attack roles the Skywarrior, originally designed as a high-altitude strategic bomber, found limited success as the aircraft was not well-suited for the type of aerial warfare found in the Vietnam theatre which often required aggressive, low-altitude flying and dive-bombing techniques. Flying missions at low altitudes was simply not an option for the aircraft as by 1967 NVN AAA defenses had expanded greatly throughout the country and become very effective, making the subsonic aircraft highly susceptible. Flying at higher altitudes opened them up to SAMs, which ringed most high-value targets in North Vietnam. Consequently, more modern and capable low-level attack and fighter aircraft, such as the A-6 and F-4, supplanted the Skywarrior's bombing capabilities. This proved to be fortuitous, as inspections in the late 1960s found that a number of A-3s had developed stress cracks at the wing roots, resulting in a wing life extension program.

While serving a diversity of missions such as attack, reconnaissance, and electronic warfare, the role of aerial tanker was by far the Skywarrior's most recognized and appreciated duty during its service in SEA. Tanker Whales played an integral and essential role in combat operations, and as such were the first aircraft to launch and the last to recover. As it was customary for three out of four air wings on Yankee Station to have A-3 tankers attached, the unique green anti-collision lights of the Whale were a constant and welcome presence over the gulf, day or night, and often in foul weather conditions.

A Skywarrior tanker could carry over fifteen tons of fuel, which was distributed across five cells through a series of pumps and valves. There was a cell in each wing (4,413 pounds/649 gallons each), a forward cell (8,255 pounds/1,214 gallons), an aft cell (12,981pounds/1,909 gallons), and an auxiliary cell (5,000 pounds/735 gallons) in the bomb bay that was used to fuel receivers. After takeoff fuel in the wing cells was transferred to the forward cell, which was then transferred to the aft and auxiliary cells as needed. Fuel in the aft cell, which fed the engines, could not be transferred to another cell or dumped, only burned. Fortunately, the critical task of maintaining weight and balance between the fore and aft cells was handled by an innovative automatic system that kept them within prescribed limits to maintain balance and center of gravity.

Aside from unique green anti-collision lights, an A-3 tanker was also readily identifiable by the refueling hardware. A drogue basket was attached to a reeled 55-foot refueling hose that extended from a teardrop-shaped fairing midship on the belly of the fuselage. The tanker package was controlled by the right-seat NFO, who had a panel on his right-hand console, and among the controls were a hose in/out switch, a resettable fuel counter that counted down in gallons transferred, and rotary activation switch. There were also two lights on his console, green and amber. These lights were replicated on the belly of the aircraft, above the basket and hose fairing where the receiver could easily view them. There was also a guillotine switch on the console to sever the hose and basket should they become a hazard in flight.

Cockpit view of a Whale taking on fuel from another A-3 over the Tonkin Gulf.

When not refueling both amber and green lights would be off. When an aircraft needing fuel rendezvoused with a Whale the receiving pilot would state the amount he required, and the NFO would set the fuel counter and trail the hose. When the hose and basket reached "full trail" the amber light would illuminate, signaling the receiver to hook up. The NFO would then turn on the activation switch and the receiver would push the basket forward 5-10 feet which would start the transfer of fuel, which turned off the amber light and illuminated the green light. If the receiver disengaged before the transfer was complete the green light would go off and the amber light would again illuminate, signaling the receiver to push back in. When the timer counted down to zero the

NFO would turn off the activation switch, which would turn off the green light, signaling the receiver to disengage. If another receiver was not ready to take on fuel the hose and basket would be reeled in.

The most common tanker sorties for Skywarriors were for aircraft returning from missions that needed fuel before recovering aboard a carrier, and to a lesser extent refueling aircraft catapulting from the carrier and needing to top-off before going feet dry on their missions. This was especially the case with the gas-guzzling F-4, which was a prodigious consumer of fuel. Phantoms were the Whale's most common customer.

Whales also commonly refueled one another, with it routine that a tanker Skywarrior coming on station to take excess fuel from the tanker they were relieving. This was not only to consolidate fuel but also to ensure the tankers going on and off-station were below maximum takeoff and landing weights. When taking on fuel from another tanker the Skywarrior pilot had a switch on his left-hand fuel control panel marked "IFR" for In-Flight Refueling. If this was selected incoming fuel would be distributed across all five fuel cells. If not selected the incoming fuel would flow into the forward cell only. If a tanker going off-station was carrying a large amount of fuel the oncoming tanker would take off with a minimal amount of JP-5, just enough to get them to a rendezvous point with the tanker going off-station to consolidate.

There were numerous challenges for tankers operating in the Tonkin Gulf, with the weather often playing a significant role. Refueling a receiver while in the turbulent reaches of thunderstorms and other foul weather that often plagued the region made the task difficult. Adding in reduced visibility made the task downright dangerous. At an airspeed of 250 knots, the 55 feet of hose that existed between the tanker and receiver during refueling could be closed in the blink of an eye.

Not all of the challenges A-3 tankers faced were weather-related. Some Whale customers returning from combat sorties had serious battle damage from AAA and SAMs, which often resulted in leaking fuel. In some instances the fuel leaked as fast as it could be replenished, necessitating the aircraft stay connected to the Whale until on final approach for landing on a carrier, or close to another boat for ditching. Hydraulic and electrical systems were also common sources of battle damage, impacting aircraft controllability, avionics, and communications. This made aerial refueling all the more challenging, and hazardous.

There were occasions when an aircraft returning to the carrier had crewmen with injuries, and any delays in refueling and trapping aboard could prove deadly. These injuries were not always physical in nature. An A-3 NFO from *Kitty Hawk* recalls a 1967 nighttime tanker mission when the pilot of an A-4 with battle damage egressed from North Vietnam and needed fuel. He was in a highly agitated mental state to the point of having difficulty hooking up to the Whale. On one attempt he collided mid-air with the tanker before successfully hooking up. Once the fuel was delivered the Skyhawk was cleared for an immediate landing. Luckily both aircraft maintained airworthiness and were able to safely trap aboard without further incident.

To deal with aircraft critically low on fuel attempting to trap aboard a carrier, Skywarrior tanker pilots developed the "Hawking" maneuver. This involved diving into the final approach leg with basket and hose trailed, positioning themselves a few hundred yards behind and above an aircraft low on fuel attempting to trap. This was known as "Trick or Treat". If the aircraft successfully landed it was a trick, and the Whale would continue, overflying the carrier before serving up the next customer. If the aircraft boltered or otherwise failed to land the pilot would clean up the airplane, retract the landing gear, and accelerate to 250 knots. He would then plug into the tanker which, if the timing was right, would be directly ahead. The aircraft would then take on "two grand," or 2,000 pounds (294 gallons) of jet fuel from the Whale, a treat. This was a quick process as tanker Skywarriors could transfer JP-5 at a rate of 2,856 pounds (420 gallons) per minute. This amount of fuel normally sufficed for several additional landing attempts.

There were also times carrier pilots would experience what was referred to as a "night in the barrel". It was a term used when, due to inclement weather, fouled deck, or other factors aircraft were delayed or otherwise experienced difficulty trapping back aboard the carrier. During these periods Whales would rarely have a moment of inaction as numerous aircraft would be egressing from combat missions and orbiting the carriers waiting to land, some in low-fuel states. A dark night with dozens of aircraft orbiting around carriers awaiting their turn to trap raised the danger level considerably for all involved, leaving a very thin margin of error.

Once combat aircraft began their treks into Vietnam Whales would commonly loiter at pre-defined stations, flying off the coastline feet wet. They would normally maintain a minimum distance, typically 22-25 miles from the nearest known SAM location, keeping them out of range of the deadly missiles. For large Alpha strikes that required radio and radar jamming multiple EKA-3Bs could be stationed as little as 12 miles from the coast at different altitudes and bearings to ensure coverage. A Red Crown PIRAZ boat or E-2 Hawkeye flying over the gulf would assist, giving the Whales navigational guidance which kept them clear of threatening areas while alerting them to any potential dangers. Aircraft that did not have enough fuel to reach a carrier due to battle damage or other factors only need to make it to a tanker, taking on enough JP-5 to make it back to the boat, or to an alternate landing or ditching site. There were also seldom occasions when EKA-3Bs would fly to USAF bases in Thailand, commonly Udorn and Ubon, and direct their ECM into Laos, Cambodia, and North Vietnam on the return trip to FASU Da Nang or carriers on Yankee Station.

The Skywarrior's first foray into combat in SEA came on 5 August 1964, during *Operation Pierce Arrow* strikes on NVN port facilities and POL sites in retaliation for the Gulf of Tonkin Incident, a motor boat attack on the Navy destroyer *Maddox* and the alleged second attack on *Maddox* as well as the destroyer *Turner Joy*. Whales attached to VAH-4 *Four Runners* on *Tico* and VAH-10 *Vikings* on *Constellation* provided tanker support to carrier-based attack and fighter aircraft during *Pierce Arrow*, which was the first major Navy strike of the conflict. Whale squadrons VAH-2 on *Ranger* and VAH-8 *Fireballers* on *Midway* also participated in these early missions.

The first Skywarrior loss in theatre was operational in nature and occurred nearly five months after the *Pierce Arrow* strikes. On 27 December 1964 VAH-4 Det. Lima attached to CVW-21 on *Hancock* lost a Skywarrior (BuNo. 142250) during a low-level training exercise in the South China Sea. During the flight, an adjacent A-3 piloted by Lieutenant Carroll Crain noticed a fuel leak coming from the starboard engine. Without warning the engine exploded and the aircraft caught fire. Three crewmen successfully egressed and were rescued. Tragically, ATN2 George Coubrough lost his life.

A second operational loss occurred on 24 February 1965 when an A-3B (BuNo. 147664) tanker of VAH-2 attached to *Coral Sea* suffered a malfunction as a result of an air leak within the fuel transfer system. The pilot climbed to 13,000 feet and the four crew bailed out, with helicopters from *Coral Sea* and USS *Yorktown* (CVA-10) taking part in the SAR effort. While three of the crewmen were rescued, BMC Dwight Frakes drowned.

View of *Constellation* as A-3 flies overhead and prepares for landing aboard. Another Whale is seen taxiing to a bow catapult.

This incident was followed by a third operational loss on 25 May with an A-3B (BuNo. 138947) of VAH-4 Det. Golf attached to *Oriskany*. While being launched for a day combat mission the aircraft suffered a failure of the catapult bridle hook which ripped off the nose gear. The bad cat shot resulted in the aircraft going over the bow of the carrier at low speed, and into the South China Sea. This failure could have been the result of undetected damage from a gear-up landing on 11 January 1962. Fortunately, all four crewmen survived. Unfortunately, this was the first of six catapult-related incidents the Skywarrior would experience during its service in the Vietnam conflict.

The fourth operational loss, also the result of a catapult mishap, occurred on 1 April 1966 while *Enterprise* was steaming 45 miles southwest of Hainan Island. A VAH-4 Det. Mike A-3B tanker (BuNo. 142665) suffered a nose wheel collapse on launch, the result of a broken catapult cable. The aircraft went over the bow at low speed and plunged into the water. Pilot Commander William Grayson and ADJ2 Melvin Krech were killed in the incident. NFO Lieutenant (junior grade) William Kohlrusch was pulled from the water by a UH-2 Seasprite plane guard helicopter of HC-1. Sadly, he died of his wounds in sick bay shortly thereafter.

The first combat loss of an A-3 occurred on 12 April 1966 when a KA-3B (BuNo. 142653) call sign "Hollygreen 3" of VAH-4 Det. Charlie being ferried from NAS Cubi Point to *Kitty Hawk* disappeared en route. A rescue effort from *Kitty Hawk*, *Enterprise*, and the USAF failed to find any trace of the aircraft or crew. Pilot Lieutenant Commander William Glasson, NFO Lieutenant (junior grade) Larry Jordan, PRCS Ken Pugh, and ATCS Reuben Harris were killed. A communist mouthpiece in Peking (now Beijing) subsequently claimed a People's Liberation Air Force unit shot down the aircraft. It was then confirmed through diplomatic sources that the KA-3B was indeed downed by Chinese MiGs over the Luichow Peninsula, which separates Hainan Island from mainland China via the narrow Hainan Strait, roughly 200 miles northeast of Yankee Station. In December 1975 cremains purported to be of Pugh were turned over by the Chinese. They were later interred at Fort Rosecrans National Cemetery in San Diego.

It is a perplexing incident due to it being unlikely an experienced crew would have, of their own accord, flown so far off-course and directly into Chinese airspace. A theory from within the air wing alleges a scenario where the aircraft suffered an oxygen system malfunction or decompression, resulting in hypoxia and incapacitation of the crew. Another theory is a possibility the aircraft was lured into Chinese airspace via electronic warfare.

Although no EA-3B airframes were lost in the Vietnam conflict, an unfortunate incident on 26 May 1966 resulted in the tragic loss of four crewmen. On that night an EA-3B of VQ-2 *Batmen* TDY to VQ-1 launched from NAS Cubi Point on an urgent mission to North Vietnam. While climbing out the Whale (BuNo. 142257) flew into a typhoon, resulting in a double-engine failure at 15,000-18,000 feet. With no engine power, the instruments began to fail, and the aircraft entered a steep descent. The emergency RAT was deployed in a bid to restore some power, followed by several attempts to re-light the engines. When this failed the bail-out order was given by pilot Lieutenant Commander Caswell. Three enlisted intelligence specialists—ATC Joseph Aubin, ATR3 Richard Hunt, and ATR3 Richard Stocker, along with their NFO supervisor Lieutenant Walter Linzy, bailed out over the South China Sea. As the aircraft descended below 8,000 feet Caswell was able to re-light both engines, at the same time plane captain Petty Officer Pickett was about to exit the aircraft. Luckily, he was stopped by Caswell, who pulled him back into the cockpit from the brink of the escape hatch.

The EA-3B was able to return to Cubi Point and land, however, the four crewmen who bailed out lost their lives. Only the body of Stocker was recovered, five days later by a destroyer. It was estimated he had died just eight hours prior. The only sign of the other three men was the

"Mae West" life preserver of Linzy, which had a note scrawled on it that read "We are in the water and OK". The aircraft was subsequently repaired and returned to service.

EA-3B on final approach for landing aboard *Kitty Hawk*.

Just over two weeks later an RA-3B (BuNo. 144842) was lost in combat. On 13 June, the recon Skywarrior of VAP-61 *World Recorders* call sign "Quiz Show 907," attached to *Hancock*, launched from NAS Cubi Point on a low-level night reconnaissance mission over Ha Tinh Province, North Vietnam. Another aircraft in the vicinity remarked observing a bright flash of light over the mouth of the Gia Hoi River, presumably the RA-3B getting shot down by AAA. No radio communications or locater beacons were detected, and no traces of the aircraft or three crewmen onboard—pilot Lieutenant Commander John Glanville, NFO Lieutenant (junior grade) George Gierak, and PTC Bennie Richard Lambton, were ever found. Though declared KIA, their ultimate fates remain unknown to this very day.

On 2 October another catapult malfunction sent an A-3B (BuNo. 142633) of VAH-2 Det. Alpha plunging into the ocean while attempting to take off from *Coral Sea*. The bridle shed during the catapult shot as a result of a broken catapult hook. Loud mechanical noises emanated from the aircraft as it veered to the left and departed the flight deck, hanging precipitously for several minutes before falling into the water partially inverted, giving the crew time and opportunity to egress. All four crewmen survived.

Tragedy struck *Oriskany* on the morning of 26 October when multiple Mk.24 magnesium flares ignited in a storage locker, the first of three major conflagrations to occur on Navy carriers during the conflict. Of the 36 officers and 8 enlisted men who lost their lives in the fire, three were from VAH-4 Det. Golf. Commander George Farris as well as Lieutenants John Francis and James Smith died from smoke inhalation.

The air war over North Vietnam escalated in February 1967, with the Johnson Administration exerting maximum pressure on the north to halt the flow of personnel and equipment into the south, and compel peace negotiations. Part of this escalation were attacks on infrastructure within Hanoi and Haiphong, as well as the mining of select North Vietnamese waterways. Following the successful 26 February 1967 mining of the Song Ca and Song Giang rivers by A-6s of VA-35 from *Enterprise,* in the first week of March A-3s of VAH-4 Det. Charlie on *Kitty Hawk* carried out two similar river mining missions. The first mission was flown by detachment OinC Commander Charles "Chuck" Keathly, B/N Lieutenant (junior grade) McIntyre, and CPO Gordon. This mission was deemed successful, resulting in a second mining mission scheduled for the following night. The crew for the second mission were pilot Lieutenant Commander Cobb, B/N Lieutenant (junior grade) Snow, and ATR2 Lipski.

Though also deemed successful, there was a problem with the second mission. When the A-3 returned to *Kitty Hawk* it was noticed that arming wires for the mines were not attached to the bomb racks as normal. This suggested that when the mines were dropped the arming wires, instead of being pulled out as the mines were deployed which would have armed them, may have instead stayed attached to the mines. This meant the ordnance may have been deployed in safe mode, rendering them useless. Personnel insisted they had wired the mines correctly, however, no explanation could be established as to why the arming wires did not remain attached to the bomb racks as normal. At that point it was decided by Keathly that subsequent mining missions would be flown with the arming wires attached normally, however, the wires would also be wrapped around the bomb racks. This would ostensibly ensure the mines were armed and prevent them from being dropped in safe mode.

Two subsequent mining missions by Det. Charlie were scheduled for the following week, on the night of 8 March. Two A-3s, launched in separate cycles, would drop mines at the mouth of the Kien Giang River from different angles. The first Skywarrior (BuNo. 144627) call sign "Hollygreen 5" launched from *Kitty Hawk*, flown by pilot Lieutenant Commander Carroll Crain, B/N Lieutenant (junior grade) George Pawlish, and ATC Ronald Galvin. They were briefed to begin their run-in to the IP at a precise time, approaching the target at 300 knots and 300 feet above the water. This would give A-4 Skyhawks time to perform flak suppression around the target site beforehand.

Shortly after Hollygreen 5 catapulted from *Kitty Hawk*, the carrier contacted them via radio, informing them the A-4s had been delayed and to suspend their run-in to the target. After a brief and peeved radio acknowledgment from pilot Crain, the aircraft vanished near Dong Hoi. No locater beacons or radio transmissions were received from the crew.

The aircraft and men aboard disappeared without a trace. A subsequent investigation, spurred by Crain's son, revealed the aircraft may have made it feet dry, executed a hard right turn in the vicinity of the river mouth, and flew into mountainous terrain in the area. Some believe that the crew may have succeeded in deploying the mines before the aircraft disappeared. The basis of this is that when *Kitty Hawk* radioed the crew to inform them that A-4s had been delayed, it was several minutes after they were scheduled to begin their run-in to the target from the IP. Therefore they were likely very near or even past the target when this transmission was received. When contact was lost with the aircraft a decision was made to stand down from the second mission. Thus, the maritime mining missions flown the first week of March 1967 were the last time Skywarriors dropped ordnance in combat during its service life.

Excavations of the area in the decades since yielded remains of an A-3, though it has never been conclusively established as being the same Skywarrior despite no records of another Whale being lost in that area. The government of Vietnam has been reluctant to allow further recovery of the wreckage, preventing resolution for the families of the men and those who served with them.

On 31 May 1967, a unique KA-3B mission[73] took place. A massive Alpha strike had been launched from *Bon Homme Richard* that day against Kep Airfield, a major VPAF base some 35 miles northeast of Hanoi. Among the support aircraft were two KA-3Bs of VAH-4 Det. Lima and two A-4E Skyhawks of VA-212, which had "buddy system" tanker capabilities, albeit with much less capacity than the KA-3B. In addition to the four tankers, the main strike force consisted of 20 Skyhawks of VA-76 (A-4C) and VA-212 (A-4E).

Three F-4s of VF-111 from *Coral Sea* that had just refueled from a Whale. This F-4 is confirming to the A-3 that the hose and reel had fully retracted into the fairing.

The lead Whale was being flown by pilot Commander John "Red Baron" Wunsch, who acquired the nickname not only for his red hair and mustache but also for his practice of performing impromptu aerobatics in the A-3. With him in the cockpit were NFO Lieutenant (junior grade) Thomas Parnella and AOC James Riley. His wingman was pilot Lieutenant Commander Donald Alberg, along with NFO Lieutenant (junior grade) Frank Stuart and AE2 Larry Davis.

After launch Alberg and his crew topped off the Skyhawk tankers and strike aircraft, as well as the KA-3B piloted by Wunsch, who was also topping off aircraft. Alberg then returned to *Bon Homme Richard* to hot pump, a process where engines were left running while the aircraft was refueled. This bypassed time-consuming engine and systems restarts so they could return to the air quickly to meet the Skyhawks returning from the strike. While Alberg and crew were refueling Wunsch and his crew stayed on station, loitering off the coast near the egress point.

There they awaited the return of strike aircraft, while carefully staying out of range of coastal air defenses.

The strike group had not yet reached the target when AAA erupted below them. Flight leader Lieutenant Commander Arvin Chauncey, flying an A-4E (BuNo. 151113) call sign "Flying Eagle 223," was hit and he ejected feet dry, approximately 70 miles inland. A SAR effort was immediately launched, while Wunsch continued to lead the effort of refueling returning strike aircraft, particularly battle-damaged stragglers. Half of the egressing Skyhawks were diverted from returning to the carrier, instead staying to provide RESCAP for SAR forces trying to extract Chauncey. One of these Skyhawks, (BuNo. 151183) call sign "Flying Eagle 229" piloted by Lieutenant (junior grade) M. Daniels, was also hit by AAA eight miles north of Kep. Flying out to sea and unable to timely locate a tanker due to a damaged radio, he spotted a destroyer and ejected. A UH-2 Seasprite helicopter rescued him from the water.

As fate would have it, at a critical moment Wunsch's aircraft suffered a fuel transfer malfunction. This prohibited him from transferring fuel from the auxiliary tank that was located in the bomb bay, which fed receivers, to the aft fuel cell which fed the engines. Although he could still refuel aircraft from the auxiliary tank, he was unable to use this JP-5 to keep his own Skywarrior aloft. Consequently, his fuel state began to dwindle, leaving him with only 15 minutes of flight time.

Alberg and his crew returned from hot pumping aboard *Bonnie Dick*, giving Wunsch what fuel they could spare, while both Whales continued to refuel strike aircraft. The KA-3Bs began to again run dangerously low on JP-5, putting their ability to return to the carrier in jeopardy. With numerous strikes occurring concurrently over North Vietnam and Laos, tanker Whales were stretched thin, resulting in the prioritizing of damaged and SAR aircraft.

This left both Alberg and Wunsch in a perilous predicament. When ditching seemed inevitable a USAF Boeing KC-135 Stratotanker was diverted to assist. Though the Air Force uses a boom system for inflight refueling, the massive tanker was also equipped with the hose and basket system used by the Navy. By that point Alberg did not have enough fuel to reach the KC-135, so Wunsch gave him what little he could spare. The two Whales rendezvoused with the Stratotanker, with each only having three minutes of fuel remaining before their engines began winding down. Wunsch plugged in first, taking on 2,300 pounds (338 gallons) of fuel, followed by Alberg.

During this time, a pair of F-8E Crusaders egressed feet wet, both in critical fuel states. Wunsch refueled one F-8 while the other plugged into Alberg, who was still taking on fuel from the KC-135, creating an impromptu daisy chain.[74] With their critical work done both Whales returned to *Bon Homme Richard* and trapped back aboard the carrier. During the mission Wunsch and Alberg refueled more than 41 aircraft, transferring some 80,000 pounds (11,764 gallons) of JP-5. Eleven of these aircraft were critically low and fuel and would have not made it back if not for the efforts of the Whales. Wunsch, Alberg, Parnella, and Riley were awarded Air Medals for their heroic efforts. The SAR operation, which lasted ninety minutes, was unsuccessful. Chauncey was captured and spent the duration of the conflict as a POW. He was released during *Operation Homecoming* in 1973.

This was not the only notable event involving the indomitable Wunsch. One day as he was waiting to trap aboard *Bonnie Dick* on Yankee Station a Soviet trawler began harassing the carrier, hazardously disrupting aircraft recovery. He took matters into his own hands, flying the KA-3B low over the offending vessel while simultaneously dumping fuel on them. The Soviet trawler, drenched in flammable JP-5, ceased their harassment. They instead frantically hosed their boat down before an errant spark doomed them to a grisly, charred fate.[75]

Though prohibited from going feet dry over North Vietnam, on 18 July 1967 a Whale crew defied this standing order to assist the pilot of an F-8. On that morning Lieutenant Commander Dick "Brown Bear" Schaffert was flying BARCAP over the gulf in an F-8C (BuNo. 146991) call sign "Old Nick 101" of VF-111 from *Intrepid*. Over the radio, he heard that an A-4 piloted by Lieutenant Commander Dick Hartman (BuNo. 151986) of VA-164 *Ghostriders* from *Oriskany* had been shot down by 37mm AAA while on a mission to the Co Trai railway and road bridge, and a rescue operation was being commenced. Schaffert went feet dry to assist and en route heard a second A-4 (BuNo. 151175), piloted by Lieutenant (junior grade) Larry Duthie, also of VA-164, had similarly been hit by 37mm AAA while flying cover over Hartman's position. Duthie tried to make it feet wet but was forced to eject near Nam Dinh, about 12 miles from Hartman. Both men were down in an area of North Vietnam with a high concentration of enemy troops and air defenses, making a rescue effort especially risky and dangerous.

At this same time Lieutenant Commander Tom Maxwell of VAH-4 Det. Golf and his crew of NFO Lieutenant (junior grade) Jim Vanderhoek and ADJ1 Petty Officer Bill Shelton were aboard *Oriskany*, getting hot-pumped to take off for their second sortie of the day. They were in the cockpit of their KA-3B (BuNo. 142655) when they were informed of the SAR effort for Hartman and Duthie. After finishing refueling they took off from *Oriskany*, orbiting about 20 miles offshore, out of range of coastal air defenses.

As on-scene commander of the rescue effort, Schaffert had burned through a large amount of fuel. With A-1 Skyraiders and other aircraft arriving on station, he turned his Crusader east towards the gulf, however, he did not have enough fuel to reach the coastline, let alone a carrier. He broadcast a mayday on the radio, which was heard by Maxwell and his crew. Knowing full well the mortal danger Schaffert faced if he was forced to eject over that area of North Vietnam, a quick consensus of the three men on the Skywarrior ended with them turning west and going feet dry.

Guided by Red Crown PIRAZ in the gulf, Maxwell and his crew flew 30-40 miles inland, into a hornet's nest of SAMs and AAA. They found Schaffert and began maneuvering to deliver fuel. As he plugged into the Whale flak bursts exploded around the two aircraft while tracers streaked by. They stayed connected until just before they reached the coastline, with the Crusader taking over 1,200 lbs. (176 gallons) of fuel, enough to reach *Oriskany*. Both aircraft trapped safely back aboard the carrier without incident.

Duthie[76] was rescued later that day by a USAF HH-3E Jolly Green Giant call sign "Jolly Green 37" from *Lima Site 38*, one of several clandestine American *Lima* bases in eastern Laos. The pilot of Jolly Green 37, Major Glen York, was awarded the Air Force Cross for the gallant rescue. The effort to extract Hartman continued into the next day and after losing three aircraft, including a Navy SH-3A Sea King SAR helicopter call sign "Big Mother" with all five men aboard KIA, the rescue was called off. On 6 March 1974, Hartman's remains were repatriated, with the North Vietnamese claiming he died in a hospital on 22 July 1967, three days after he was shot down.

Though deemed heroic and celebrated throughout the Skywarrior community, as their act of going feet dry violated a standing order, the crew received no formal recognition for their valiant, selfless, and lifesaving deed.[77]

EKA-3Bs in formation over the Tonkin Gulf.

A KA-3B tanker (BuNo. 142658) of VAH-4 Det. Golf from *Oriskany* was lost to the notorious weather of the Tonkin Gulf on 28 July 1967, some 150 miles northeast of Da Nang. While flying at high altitude a torrential downpour resulted in double-engine failure. While the crew had time to egress from the aircraft only pilot Lieutenant Commander Michael Kavanaugh survived, rescued by a USAF HH-3E Jolly Green Giant from the 37th ARRS. B/N Ensign Bruce Patterson and AE2 Charles Hardie lost their lives. The aircraft had just been converted to an KA-3B at NAS Alameda the previous month.

Just one month later, on 25 August 1967, an RA-3B (BuNo. 144835) of VAP-62 TDY to VAP-61 call sign "Quiz Show 9" based at NAS Agana and operating from FASU Da Nang, disappeared while on an IR night road recce mission. As with the RA-3B loss of Quiz Show 907 on 13 June 1966, no radio communication or locater beacons were detected, and no trace of the aircraft or crew were ever found. Pilot Commander Edward Jacobs, NFO Lieutenant (junior grade) James Zavocky, and ADJ2 Ronald Claire remain missing, their fates unknown. This particular Skywarrior had previously sustained AAA damage on 15 August that resulted in a punctured wing fuel cell, however, the aircraft was able to safely return to base and the damage subsequently repaired.

As 1967 drew to a conclusion, American military aircraft losses surpassed the previous year's startling totals. On 14 October an RA-3B (BuNo. 144844) call sign "Quiz Show" of VAP-61 from NAS Cubi Point was flying a daytime coastal reconnaissance mission when it was hit by AAA 15 miles south of Thanh Hoa. Pilot Lieutenant Commander Robert Vaughan, turned feet wet to find a Navy ship in the Tonkin Gulf. They had made it ten miles out to sea when the aircraft impacted the water near the destroyer USS *William V. Pratt* (DLG-13). Vaughan was killed in the incident and his body was not recovered while NFO Lieutenant (junior grade) Mosler and ADJ2 Shaw managed to bail out. They were rescued by a UH-2 Seasprite helicopter of HC-7 from *Pratt*.

On 21 October 1967 a KA-3B (BuNo. 142655) of VAH-4 Det. Golf was using four JATO bottles during a takeoff from Runway 25 at NAS Cubi Point on a logistics flight to *Oriskany*. FOD entered the starboard engine, causing the aircraft to veer off the runway. The pilot managed to get the Whale momentarily airborne, however, the damaged engine failed. The aircraft was ditched, with all four crewmen luckily surviving without serious injury.

Fatal catapult mishaps continued to shadow the Skywarrior's service. On 3 November 1967 a KA-3B (BuNo. 147653) of VAH-8 Det. 64 was launching from *Constellation* when, ninety feet into the shot, the catapult bridle separated from the aircraft and it went over the bow with insufficient speed to maintain flight. Pilot Lieutenant Commander Peter Krusi, NFO Lieutenant (junior grade) Hans Grauert, and ADJ2 Richard Sandifer were killed, with only the body of Sandifer recovered.

Night road recce missions were particularly dangerous for the unarmed RA-3B. Unlike many of its fighter, attack, and reconnaissance counterparts, it did not have the speed, nor agility, to evade AAA and SAMs. On New Year's Day 1968, an RA-3B (BuNo. 144847) call sign "Quiz Show 301" of VAP-61 attached to *Oriskany* flying from Cubi Point was on a night road recce mission and had just gone feet dry when it was hit by AAA. The pilot, Lieutenant Commander James Dennison, attempted to fly out to sea where the crew stood a better chance of rescue. The aircraft impacted water about 30 miles from Dong Hoi. No trace of the Skywarrior or pilot Dennison, NFO Lieutenant (junior grade) Terence Hanley, or PHCS Henry Herrin, were ever located. This VAP-61 RA-3B was the first American military aircraft lost in the Vietnam theatre in 1968, and the last Skywarrior lost in the conflict to combat causes.

On 31 January 1968, North Vietnam launched the Tet Offensive, which in March resulted in President Johnson ordering another halt to the bombing of North Vietnam. This essentially ended *Operation Rolling Thunder*, which began in April 1965. The focus of air operations became *Operation Commando Hunt*, which targeted sections of the Ho Chi Minh Trail in North Vietnam, South Vietnam, Cambodia, and Laos. Despite the ferocity of the offensive and major battles fought, the RA-3B lost on New Year's Day was the only Skywarrior lost that dark year.

After a tumultuous 1968 which deeply fractured the country and inflicted terrible damage on his administration, President Johnson chose not to run for re-election. This resulted in Richard Nixon being sworn in as President on 20 January 1969. Nixon's election marked a significant change toward American foreign policy in SEA that began to be reflected via Vietnamization. As pledged during his campaign, the number of American military personnel in the region began to steadily decline following his inauguration.

A month after Nixon took office, another Skywarrior was lost, the first loss of a Whale in theatre since 1 January 1968. On 17 February 1969, after a nighttime tanker mission, the KA-3B (BuNo. 138943) of VAH-10 Det. 43 attached to *Coral Sea* began preparation to trap aboard the carrier. The aircraft was suffering a fuel system problem and had been unable to take on JP-5 from another tanker, which had noted fuel streaming from its drogue basket. While flying a TACAN approach, pilot Lieutenant Commander Rodney Chapman was instructed to make an inbound turn towards the ship. The aircraft failed to make the turn, continued outbound, and did not respond to subsequent radio calls, eventually disappearing from radar. An extensive search found no sign of the aircraft or pilot Chapman, AMS1 Stanley Jerome, or AO1 Eddie Schimmels. The cause of the tragedy remains unknown, with speculation of fuel exhaustion or pilot fatigue.

Nearly six months later a systems malfunction caused the loss of an RA-3B (BuNo. 144826) of VAP-61 based at Cubi Point TDY to Da Nang. On 8 August 1969, while the crew were concentrating on solving a fuel transfer, problem both engines flamed out, and attempts to re-light were unsuccessful. Pilot Commander James Berry, NFO Lieutenant (junior grade) Christopher Overton, AQBAN Jon White, and civilian Cecil Brock bailed out over Laos, giving them scant odds of survival.

While Overton and White landed terra firma, the parachutes of Berry and Brock got hung up in trees short of the ground. Luckily for them, a USAF HH-53B Super Jolly Green Giant of the 40th ARRS and *Air America* (CIA) UH-34 Choctaw were in the area. They were plucked from the unforgiving Laotian jungle before the murderous Pathet Lao or other hostile elements could capture them.

As President Nixon's ongoing Vietnamization process wound down American presence in theatre, air losses began to decline, though every loss was painfully felt by a military that had become both weary and cynical of the prolonged conflict. In the late afternoon of 16 May 1970, an EKA-3B (BuNo. 142657) of VAQ-135 that had undergone maintenance at NAS Cubi Point was being ferried back to Da Nang during a period of intense thunderstorms over the Tonkin Gulf.

The aircraft was piloted by squadron XO Commander Richard Skeen, along with NFO Lieutenant Commander Eugene "Gene" McNally and ADCS Edwin Conner. They were initially supposed to transport a VIP, reportedly an admiral, to *Coral Sea* on Yankee Station before continuing to Da Nang. However, the admiral did not make the flight. The crew still planned to trap aboard *Coral Sea*, however they were instructed to continue to Da Nang.

When the aircraft failed to materialize a SAR effort was launched. The destroyer USS *Waddell* (DDG-24) eventually found a small amount of wreckage in the waters off the coast of South Vietnam, along with the body of McNally. The remains of Skeen and Conner were not recovered. McNally, a native of Montana, did not normally fly with Skeen, but had gone on the trip to Cubi Point to visit his wife and two children, who were at the time living in the Philippines.

The most likely cause of the loss was convective weather. Some within the squadron believe there is a possibility the Air Turbine Motors (ATM), which provided power to the hydraulic actuators and other electrical components on the aircraft, may have been knocked off-line due to turbulence or attitude upset. The ATMs had nozzle vanes that controlled the flow of bleed air to the turbine motors. Negative g-forces lasting 15 seconds or longer could cause these vanes to close, rendering the ATMs inoperative. If this occurred the pilot would pull two handles located directly under the yoke. One handle was for ailerons, the other for elevator and rudder. Pulling these handles allowed control of the Skywarrior minus hydraulics, although this would have resulted in the aircraft becoming extremely difficult to control at a most inopportune time. The process of resetting the ATMs in an EKA-3B called for the cockpit to be depressurized and a crew member enter the companionway to the bomb bay to manually reset them.

Less than two months later, on 4 July 1970, an EKA-3B (BuNo. 142400) of VAQ-132 *Scorpions* attempting a night trap on *America* boltered and began to climb out. The aircraft's drogue chute unexpectedly deployed, the result of a malfunction experienced during an earlier landing at Da Nang when the chute failed to deploy. The air boss immediately informed the pilot, Lieutenant William "Splash" Whitlow, Jr. of the chute being deployed, which should have resulted in him throwing a switch located under the magnetic (wet) compass to jettison it. However, while troubleshooting the malfunction earlier that day at Da Nang, a procedure had been performed which effectively disabled the jettison function. Due to the tremendous drag, Whitlow began to lose control at low airspeed and altitude. The aircraft was ditched in the water, with all three crewmen rescued. It was this incident that resulted in Whitlow acquiring the Splash nickname.

It was nearly a year later, on 18 June 1971, when an EKA-3B (BuNo. 147649) call sign "Waste Basket 614" of VAQ-130 *Zappers* Det. 3, attached to CVW-19 on *Oriskany* flying from FASU Da Nang, was lost while on a tanker mission. The aircraft had just completed refueling two F-8 Crusader fighters when pilot Lieutenant John "Johnny" Painter executed a maneuver that resulted in a loss of control. The Whale dived nose-first into the ocean from an altitude of 20,000 feet, some 200 miles north of Da Nang. An extensive search for Painter, NFO Lieutenant (junior grade) Raymond DeBlasio, and ADJ2 Barry Bidwell failed to locate any remains.

Painter, who was the operations officer of the squadron, was an experienced scuba diver as well as a skydiver, with several hundred jumps to his credit. Known for his daredevil exploits, tellingly his favorite song was "The Ballad of the Green Berets". DeBlasio was known among his squadron mates as "Crazy Ray," a name bestowed due to his rather eccentric personality. However, those who served with him recall that, personality notwithstanding, he was a highly talented NFO. The death of 23-year-old Petty Officer Bidwell came as a tremendous shock to his family. They were unaware he had volunteered as aircrew, and therefore had no inkling he was flying combat support missions.

Four years after the Tet Offensive and the subsequent end of *Operation Rolling Thunder*, on the afternoon of 30 March 1972, North Vietnam launched the Easter Offensive, with major PAVN/VC assaults targeting the DMZ as well as the strategically important areas of An Loc, Kon Tum, and Quang Tri. Though a spring offensive was expected by South Vietnam and the United States, both the size and ferocity of the operation were underestimated. In echoes of the Tet Offensive that began in January 1968, the surprise attack by the North Vietnamese resulted in significant inroads into the south across multiple fronts.

American ground forces in South Vietnam in the spring of 1972 were a fraction of what they were in January 1968 when the Tet Offensive began. Vietnamization had resulted in less than 70,000 military personnel in theatre, of which fewer than 10,000 were combat troops, and less than 100 combat aircraft. However, at the time the offensive began the Navy had two carriers on Yankee Station, *Coral Sea*, and *Hancock*, with approximately 140 aircraft. *Constellation* had taken a scheduled break from line operations on 26 March and docked in Yokosuka, Japan on 31 March. After only four days the carrier left port and returned to line operations on Yankee Station. *Kitty Hawk* was in-port in Subic Bay, departing on 31 March to return to the line. *Midway* cut their west coast ORE/ORI exercises short and hastily departed NAS Alameda on 16 April. Shortly after *Midway's* departure, the carriers USS *Saratoga* (CVA-60) and *America* would steam post-haste for the Tonkin Gulf on 8 May and 1 July, respectively. *Oriskany*[78] also sped up its seventh scheduled Vietnam deployment, departing Alameda on 5 June and arriving in Subic Bay on 21 June. This rapid build-up would see the largest number of carriers ever amassed on Yankee Station.

With ground counter-attacks by South Vietnamese forces largely ineffective, raw air power eventually inflicted heavy losses on enemy forces, blunting the offensive. *Operation Freedom Train*, a month-long bombardment campaign that brought B-52 bombers back into the foray, launched on 5 April. This was followed by *Operation Linebacker*, which began on 9 May. *Operation Linebacker* lasted through 23 October, when the North Vietnamese agreed to return to the bargaining table. Their stubborn recalcitrance resulted in the launch of *Operation Linebacker II* on 18 December, which was the last major air offensive of the conflict. The North Vietnamese suffered high personnel losses and extensive material damages during *Linebacker* operations, which once again pushed them to diplomacy in Paris. However, they were able to achieve several strategic objectives during the offensive, which put them in an advantageous position for the ongoing peace negotiations.

Les Parker, Steve Paine and Keith Christophersen. Paine, along with his wife and three children, lost their lives in a 1995 aircraft accident in Florida.

On 21 January 1973, the twentieth Skywarrior loss in theatre occurred. On that night an EKA-3B (BuNo. 142634) of VAQ-130 Det. 4 was launching from *Ranger*, which was steaming some 100 miles east of Vinh. Sixty feet into the shot the catapult hook separated from the bridle, resulting in sparks and an explosion coming from the starboard engine. The aircraft, which was near 73,000 pounds MTOW, left the deck at low speed, nosed into the sea, and sank.

Pilot Lieutenant Commander Charles "Les" Parker, who was assistant OinC of the detachment, NFO Lieutenant (junior grade) Keith Christophersen, and AT2 Richard Wiehr lost their lives. Though neither the aircraft nor bodies of the crew were recovered, a subsequent investigation theorized that the bridle had been mispositioned on one of the catapult hooks, contributing to its shedding. The sparks emanating from the engine and explosion were most likely due to ingesting FOD from a catapult hook.

Lieutenant (junior grade) Alec Schmidt was an NFO with the detachment at the time of the loss. He had considered riding with the crew in the fourth seat, but at the last moment fatefully decided against it. He vividly recalls the incident:

> I attended the preflight briefing in ready room two, which was located on the O-2 level just below the forward port catapult. CVW-2 was just starting midnight-noon flight operations and I believe this was the first launch of this new line period. The A-3 was usually the first aircraft to launch, typically from the waist catapult.
>
> In ready room two there was a characteristic sound made when the catapult fired. When Parker, Christophersen, and Wiehr launched, the sound made was quite atypical, signaling something was not right. Almost immediately after that, an announcement came over 1MC of a plane in the water. It had shed the bridle after only about 60 feet of travel and they nosed off the waist catapult into the water.

Following his loss, the devastated wife of pilot Les Parker, then a widow with a seven-year-old daughter, filed a lawsuit against McDonnell Douglas. The action alleged design and material deficiencies, along with a lack of ejection seats, led to her husband's tragic and untimely death. McDonnell Douglas initially offered to settle the matter out of court. However, that offer was ultimately withdrawn and the lawsuit unsuccessful. Those who served with Christophersen remember him as a consummate professional as well as a quiet, and private person. He hailed from Minnesota, and would often reminisce with his shipmates about his love of wilderness canoeing on the many lakes in the state. At the time he deployed on *Ranger* in November 1972, his wife had very recently given birth to a daughter. For Petty Officer Wiehr, also a native of Minnesota, it was his first combat mission in an A-3.

On final approach for landing aboard *Coral Sea*.

This EKA-3B would be the last Skywarrior lost in the conflict. Six days later, on 27 January, the Paris Peace Accords were signed, formally and finally ending the air war over North Vietnam. Of the 20 A-3s lost in eight years and five months of combat in the Vietnam theatre, six were lost to enemy action, four of those being reconnaissance RA-3Bs of VAP-61. The remaining fourteen aircraft were lost to operational accidents, six of them involving catapult mishaps. The year 1967 was the most costly for the United States in terms of aircraft losses, and as such the six Skywarriors lost that year were the apex of annual Vietnam theatre A-3 losses.[79]

By the cessation of hostilities in 1973 the Skywarrior, hitherto a seemingly unremarkable support aircraft and one of many that had served in SEA, had achieved heroic status amongst thousands of naval aviators who had flown from carriers on Dixie and Yankee Stations during the long engagement. Along with the tanker capabilities of the A-3B and KA-3B, the EA-3B had provided valuable intelligence of NVN air defenses while the RA-3B rooted out and exposed a large number of enemy elements, as well as providing important reconnaissance imagery that produced highly detailed maps of the battlefields. The EKA-3B had saved countless aircraft and lives not only as a tanker but also from its ability to jam VPAF radio communications and radars of minacious SAMs. It has been estimated that as many as 700 Navy and Marine Corps aircraft were saved by Whales during the conflict, along with the lives onboard. However, there was a price to pay well beyond the loss of replaceable aircraft. A total of 39 Skywarrior crewmen lost their lives during the conflict—17 killed in action in addition to 22 killed in operational incidents.

The successes of the Skywarrior in the Vietnam theatre are owed not only to the aircraft and those who designed and built it, but also to those intrepid men who flew them in combat. One such man is Captain David Mason, who entered the Navy as a NAVCAD and had completed flight training with VT-21 in Kingsville, Texas in October 1965, and was awaiting his next set of orders. Mason was expecting assignment to a fighter, or attack squadron, and was therefore surprised when the CO of the training squadron called him into his office and congratulated him on being one of only two students selected to fly the A-3. He was not familiar with the aircraft, and had to consult a book to identify and learn about it. After advanced instrument training in A-4s at NAS Miramar. he proceeded to RAG squadron VAH-123 at NAS Whidbey Island.

As the A-3 was classified as a bomber, much of the curricula in Whidbey revolved around that capability. In addition to training in day and night refueling, the pilots also trained in dive, "lay down" and "pop up" bombing, nuclear munitions delivery, as well as maritime mining. David also began training as an LSO[80] for future carrier deployments. By the time he completed RAG training in June 1966, he was the only ensign qualified for nuclear munitions delivery, and would later become the first ensign qualified as an A-3 aircraft commander. He was then assigned to VAH-4 Det. Charlie, where he was the most junior officer in the detachment. After ORE/ORI off the coast of California, the detachment deployed on 5 November 1966 with CVW-11 aboard *Kitty Hawk*, the second Vietnam deployment for the carrier, and the first for Det. Charlie. David made Lieutenant (junior grade) just before deploying, and was also the scheduling officer of the detachment that cruise. His additional duty as an LSO allowed him to interact with pilots from across the air wing:

I would grade the pilot's landings, then go down to the squadron ready room and debrief them. After a few months of this, I knew just about every pilot from every squadron, and formed close friendships with many of them. One of the men I got to know was VA-85 *Black Falcons* A-6 pilot Lieutenant (junior grade) Lewis Williams, who along with his B/N Lieutenant (junior grade) Michael Christian, was shot down on 24 April 1967. Both became POWs. Another man I got to know well was RVAH-13 RA-5C Vigilante pilot Charlie Putnam, who was shot down along with his RAN Frank Prendergast on 9 March 1967, resulting in one of the most memorable rescues of the conflict. Charlie did not make it out, however Frank, who was also a good friend of mine and one hell of a poker player, made it back after battling a couple of NVA regulars who attempted to capture him on the beach.

Before Charlie was lost I remember that when we steamed into port for a break from the line, he would go to the O-Club and buy 100 Stinger rounds in honor of every man that had been shot down during the preceding line period. The first break we had after he was lost, we took a collection and bought 1,000 Stingers in his honor.

Det. Charlie was billeted onboard *Kitty Hawk* that deployment, and though there were five A-3s in the detachment, only three were kept on the carrier at any time, with the remaining two kept at NAS Cubi Point. The aircraft were routinely swapped between these two locations for maintenance and other purposes. After stops in Hawaii, Japan, and

Subic Bay the carrier arrived on Yankee Station in early December 1966 and immediately began flying combat sorties. David logged over 25 missions that first month on the line.

By the time *Kitty Hawk* arrived for that deployment, the A-3 had since been removed from bombing and mining sorties. Aside from the aircraft not being particularly well-suited for these types of missions in theatre, it was also purportedly due to MACV being informed the Skywarrior did not have proper bomb sites, with B/Ns instead relying on grease pencil markings on the windscreen, or the tip of the refueling probe. David used a piece of fishing line that ran from the top of the glare shield to the top of the canopy, with a moveable metal washer to mark his lead. Though an attempt was made to install A-4 bomb sites on the glare shield of the A-3, this was found to block the pilot's view, and deemed unacceptable. Nevertheless, the squadron stayed sharp and on alert to deliver the nuclear weapons that were onboard *Kitty Hawk*. David had two pre-designated targets, both within China.

He recalls when the squadron received orders to deploy maritime mines the first week of March 1967. Though by that time the A-3 was no longer performing bombing and mining missions, OinC Commander Keathly lobbied hard for an exception. When the squadron began planning the missions one pilot, along with one B/N, both married family men, requested and were excused from flying feet dry over North Vietnam due to the high risk involved. However, David remembers those two as the exception as others in the squadron were eager for action:

> We were trained for combat, for delivering conventional and nuclear weapons on a target. We also trained in deploying maritime mines. The squadron practiced and maintained proficiency in those skills during training.

However, when we arrived in SEA we were told the only combat we would see was from the Tonkin Gulf, refueling aircraft. We were of course disappointed, therefore most of us welcomed Keathly's efforts. I remember he had originally pitched mining Haiphong Harbor, however, that plan was rejected. Instead, we started on rivers, such as what A-6s from *Enterprise* had accomplished in February.

David, along with B/N Lieutenant Scotty Meiner and AQB2 Jim Gross were the crew of the second A-3 scheduled to be launched from *Kitty Hawk* to drop maritime mines on the night of 8 March 1967, behind the first A-3 being flown by Crain, Pawlish, and Galvin. They were on the No. 1 catapult at full power for an imminent launch when they were told to stand down and secure their aircraft. It was an especially close call for Meiner, who had flipped a coin with Pawlish to determine which of them would go on the first aircraft.

He has nothing but fond memories of Crain, remembering him as a good friend whom he often played golf with; a fun-loving and outgoing individual who was a joy to be around. Unlike many of his contemporaries, Crain was not a heavy drinker. However, he nevertheless enjoyed going to the O-Club and reveling with his fellow aviators. Of his aviation skills, David praises his abilities:

> Carroll was an outstanding aviator. Of all the pilots I flew with in my military career, he was one of three that I would rate as the best stick-and-throttle men. When he flew it was highly aggressive and to the maximum. He routinely pushed the performance of the A-3 to the edges

of the envelope. For him, this was not showboating or grandstanding. As a pilot, he had the skills to back up everything he did in the cockpit.

George Pawlish was a talented NFO, as well as a quiet and laid-back family man from Bent County, Colorado. David describes him as "One of the nicest guys you would ever meet". The 1966-1967 cruise was the second deployment Crain and Pawlish flew together, and they had a close personal relationship. This is somewhat ironic considering, as David puts it, "Their individual personalities were total opposites".

Detachment OinC
Commander C.C. Keathly

A young Lieutenant (junior grade)
David Mason.

As for OinC Commander Keathly, David describes him as a colorful persona who had come from the fighter community. He was of the aggressive personality type that came with that background, and flew the A-3 in the same manner, despite the handling characteristics of the aircraft being far from that of a fighter. He was also a physically fit man with experience in the boxing ring and would not hesitate to duke it out with anyone who dared challenge him, or those who leveled what he perceived as insults. David recalls an incident at Cubi O-Club when an F-8 Crusader pilot from CVW-19 on *Tico*, well known for his brash ego and loud mouth, arrogantly insinuated to Keathly that A-3 pilots were far inferior to fighter pilots. A shoving match and brief fistfight resulted in the Crusader pilot on the floor, with Keathly standing victoriously over him.

David remembers Keathly as somewhat aloof and also having a tremendous ego. Nevertheless, he was generally considered a good OinC and considerate of those under his command. One example is granting excuses to the two married men who declined to fly risky mining missions over the beach, with no punitive action taken against them. Another example is when David had been granted leave to attend the annual Tailhook Convention taking place in San Diego. Keathly rescinded his leave, instead granting it to another detachment pilot. The reason given was the other pilot's wife was coming in from out of town, they were trying to have a child, and needed some relaxing time together to kick off the process.

With improving spring weather and the Johnson Administration's escalated efforts to push the North Vietnamese to diplomacy, the months of March and April 1967 saw a marked increase in Alpha strikes within North Vietnam. David remarks that in February he flew 13 missions, in March that number grew to 24 and in April it was 29, often with multiple missions flown per day. On Black Friday 19 May 1967, when six Navy aircraft were lost, he flew three separate missions in BuNo. 142662 over a period of four hours.

David recalls there were occasions during that hectic spring when decisions to launch a strike would be made at the last minute, on newly discovered targets of opportunity. Time was of the essence on many of these as the North Vietnamese routinely moved air defenses and other mobile assets to prevent them from being targeted. There were occasions when pilots, himself included, would be relaxing by the swimming pool in Cubi Point, or comfortably in their staterooms below deck, when they were told to get suited up. In less than an hour they would be in their aircraft, ready to fly a combat sortie.

VAH-4 Det. 63 deployed for a second Vietnam WESTPAC aboard *Kitty Hawk*, which began on 6 December 1967. On the 21st day of the following month, the NVN began an assault on Khe Sanh, and ten days later the Tet Offensive was launched. His worn logbook indicates that he flew 32 day and night missions in the month of January alone. David experienced many memorable moments during his two combat tours. He ranks consolidating with other tankers and refueling aircraft in foul weather as the most challenging, with too many close calls and near-disasters to count. He recalls that the most rewarding aspect of his tours came from his role as an LSO:

> The camaraderie I had with the men in the air wing, particularly on the first cruise, was something very special. I got to know them really well, not only as an LSO, but also personally. Along with our time together in combat and on *Kitty Hawk* we also spent many hours with each other in ports; places such as Cubi Point and Hong Kong. In the process we forged very close bonds that remain to this day. No passage of time will ever change the relationships that were forged.

Officially, A-3 pilots were prohibited from performing aerobatics. This did not, however, discourage certain Skywarrior drivers from rolling the aircraft when an opportunity presented itself. Men in the air wing fondly recall several instances when David rolled the aircraft while abeam the carrier prior to landing, providing some showmanship to his fellow sailors. When queried on this his response was "If I admit to doing that (which I did several times), am I in trouble?"

Commander Mike Murphy narrowly missed an appointment to the Air Force Academy while in his senior year of high school. When he subsequently asked a recruiter how else he could become an Air Force pilot the man's surprising but honest advice was "If you *really* want to know how to fly, join the Navy". Mike began AOCS flight training in Pensacola under the Navy AVROC program following his junior year of college, finishing after his senior year. He earned his wings in the Grumman F-9 Cougar in the spring of 1970.

While at AOCS Mike had the privilege of briefly working as an assistant to Marine Corps Major Stephen Pless. As a helicopter pilot, he had performed an improbable and heroic rescue of four stranded Marines on 19 August 1967 while assigned to Marine Observation Squadron Six (VMO-6). Pless repelled a large force of VC soldiers while using his helicopter to shield the Marines from heavy enemy fire, successfully extracting all four. He was later awarded the Congressional Medal of Honor for his gallant actions, while members of his crew received the Navy Cross. When Mike queried Pless as to how he was able to accomplish such an incredible rescue against overwhelming odds, he claimed that as he was struggling to get the damaged helicopter out of ground effect he used the rotor blades as a sort of crude guillotine to cut down charging enemy attackers. Pless lost his life as a result of a motorcycle accident in Pensacola on 20 July 1969, age 29.

Like many in his graduating class, Mike wanted to fly fighters but the Navy had different plans, based on their current needs. Upon graduation he, along with several other top performers in his class, were sent to RAG squadron VAH-123 *Professionals* at NAS Whidbey Island to begin training in the A-3. Mike remained with the RAG through August 1970, when he joined VAQ-130 Det. 3 based at NAS Alameda, flying the hybrid EKA-3B. The detachment transited to the Philippines in May 1971, attached to CVW-19 on *Oriskany* for the cruise that began on 4 June, the carrier's sixth WESTPAC of the conflict.

Det. 3 was stationed at NAS Cubi Point during that 1971 deployment, though they were billeted at, and operated primarily from, FASU Da Nang, a large airbase that in its heyday hosted units from every branch of the military as well as commercial carriers. Located on the coast only 100 miles southeast of the DMZ, the base was subject to recurring rocket and mortar attacks by enemy forces throughout the conflict, particularly during the 1968 Tet Offensive. These attacks destroyed multiple aircraft and inflicted numerous casualties. By 1972, as American presence on the base waned, the attacks were occurring several times a month, mostly in the late night and early morning hours. To minimize damage, many aircraft were in reinforced revetments, and living quarters were often surrounded by 55-gallon drums filled with concrete. Numerous underground shelters were also dug throughout the base. AC-119 Flying Boxcar gunships routinely patrolled the perimeter, with spotlights and red tracers from their Gatling guns highly visible in the night sky. A popular evening recreational activity for Mike and others was relaxing with a cocktail while watching the gunships work, the unmistakable groan of their Gatlings permeating the thick, humid tropical air.

Throughout much of the conflict, the Air Force and Marine Corps maintained a large presence at Da Nang, while the Navy's FASU on the west side of the field was diminutive by comparison. The unofficial Navy O-Club on the base, the *Red Dog Saloon*, was not much more than a dinky annex next to the officer's quarters. The club consisted of a small bar, a refrigerator, and some weathered wooden tables and chairs. The walls were covered in writing, sentiments from Navy men who had memorialized their presence. The club was named in honor of an OinC whose call sign was "Red Dog".

The meager Navy facilities were generally acceptable to personnel billeted at Da Nang, with the very important exception of beer. The most readily available domestic brand of suds was brewed in open vats, which resulted in a cavalcade of insects floating to the top when poured in a glass. Insect-free and great-tasting San Miguel beer, widely available in the Philippines, was the preferred beer of choice for FASU personnel. It therefore became somewhat of a practice for A-3 pilots based at Da Nang that when departing Cubi Point, around 110 cases of San Miguel beer in bottles (weighing 35 pounds per case) would be loaded into the bomb bay. Mike recalls carrying a load of San Miguel from Cubi Point to Da Nang one day when *Oriskany* called on him to trap aboard. He informed them he had 3,500 pounds of *high-priority cargo* bound for Da Nang, and could not get the aircraft down to max landing weight. Getting a pass, the beverage was delivered later that day, much to the delight of many sailors. Though generally kept within the FASU, San Miguel was occasionally shared with other military branches for barter, as tokens of appreciation, or to curry favors. Beer was not the only foodstuff that A-3 and other Navy aircraft routinely hauled across oceans into Da Nang. Kobe steaks and other culinary delights were regularly flown in from NAS Atsugi and other bases in Japan.

A duo of thirsty A-7 Corsairs looking to fill their tanks.

Being that Da Nang hosted units from every military branch, it was not uncommon for airmen stationed there to crossover and fly with a different branch from which they served. Sometimes this was sheerly for the experience and thrill, while other times it was a necessity for meeting currency requirements to continue receiving flight pay. Among some, it became a popular notion that a tanker sortie on a Skywarrior over the Tonkin Gulf was a safe, and easy way to accomplish this. One day in Da Nang, Mike and his two crewmen were pre-flighting their EKA-3B for a tanker and jamming mission over the Tonkin Gulf, when an Air Force flight surgeon appeared, asking to go for a ride. Though the three crew seats were filled, aircraft commander Mike offered the flight surgeon the jump seat. The Air Force doctor happily agreed, and was soon being fitted with flight gear and getting strapped into the Whale.

After taking off from Da Nang the unarmed Skywarrior refueled several aircraft over the Tonkin Gulf. The doctor was apparently enjoying his thus-routine experience when the carrier called, telling them to fly north for a "Black Track" mission, eliciting a chorus of reactions from the three crewmen. Having heard the radio transmission and ensuing chatter, the doctor inquired as to the meaning. Mike informed him they were going on a combat jamming mission southeast of Haiphong, at the time one of the most heavily defended cities in the world, ringed with abundant AAA and SAMs. It was one of the few Skywarrior sorties that counted as a combat mission, versus combat support. Mike remembers his reaction:

> The good doctor expressed his unwillingness to embark on a combat mission, but I pointed out that he was the one that asked for a ride. We completed the mission uneventfully and returned over the carrier to do some more refueling. Then the air ops officer on *Oriskany* came up on the radio and stated "Murph, you are a trap cat".

Once again confronted with unfamiliar Naval Aviation jargon, the doctor inquired as to the meaning of trap cat. Mike delivered more unwelcome news to their guest:

> I said "Doc, this is your lucky day. You not only got a combat mission, you're about to do something you can tell your grandkids about. You are going to experience an aircraft carrier landing!" He again expressed concern, nervously and sheepishly stating "I don't want to do that!"

I apologized and reminded him again that he was the one who asked for a ride. So we went into the break, which I always tried to make as colorful as possible, and trapped aboard *Oriskany*. We immediately taxied to the starboard catapult and blasted off for Da Nang without taking on any additional fuel. Our weight due to all the electronics we carried only allowed a small amount of fuel on landing, about 5,000 pounds. Catapulting off the carrier with say only 4,500 pounds put us immediately into a low-fuel situation. I explained that to the doctor, and once again he was not pleased. We rocketed from the carrier, quickly climbed to altitude, and began an idle descent into Da Nang. Once on the ground the doctor got out of the airplane, kicked off his flight gear, and stomped away. No goodbye, thank you, or have a nice day—nothing. I got the feeling he was pissed.

The Skywarrior was no stranger to the notoriously volatile weather of the Tonkin Gulf, which regardless of season, was infrequently ideal. Convective thunderstorms, particularly multiple cells within a squall line that could reach over 60,000 feet in altitude, were not uncommon over the waters of the gulf. This made a challenging environment for aircraft, particularly tankers. A competent B/N could use the onboard radar to get the aircraft around the worst of a storm, however, it was nearly impossible to dodge the effects completely. Mike remembers one night when he was the last tanker over the gulf, the "sheriff" refueling the last aircraft as they returned from combat missions:

One night I was playing sheriff, and put the ship to bed before rolling out toward Da Nang. The controller on *Oriskany* came up on the radio and said "Be advised, you have two solid lines of thunderstorms between you and Da Nang".

I replied, "Be advised we have no radar, can you vector us through the holes in the storm?" An apologetic voice came back and said "There are no holes".

I had our maintenance warrant officer "Shaky" Malone on board just for a ride, and he was to become no less shaky during the course of the evening. When we penetrated the first thunderstorm St. Elmo's fire lit up the entire airplane, extending several feet in front of all the surfaces. Then a medicine ball of St. Elmo's built up on the fuel probe, traveled down the probe and into the cockpit, disappearing into the bomb bay area. With that, Shaky just about shook out of his skin. We penetrated the second line of storms with a little less fanfare, and made it into Da Nang minus a good portion of our paint job, and a few blisters on the radome. Actually, that was a pretty normal day. The airplanes commonly became so polished from rain and hail that it was almost impossible to get paint to stick. We contemplated getting "One Hundred Thunderstorm" patches made for our flight jackets so we could brag about being "Thunderstorm Centurions".

Mike remembers another dark night when he was dispatched to assist a Whale from a sister squadron that was on a mission off the coast of North Vietnam. While flying in this area the aircraft suffered numerous navigation system failures, as well as a transponder malfunction, leaving them lost and unidentifiable in the unfriendly skies which lay uncomfortably close to Chinese airspace. They were also beginning to run low on fuel, adding urgency to the already critical situation.

He flew north, and with the help of Red Crown intercepted an unknown radar contact. As he drew closer he spotted the tell-tale green anti-collision lights of a tanker; the lost Skywarrior. Making radio contact, he advised them to start a port orbit so he could maneuver in front to give them fuel. As the two aircraft neared each other Mike noticed the navigation lights of the wayward tanker, which when approaching from behind should have been red on the left and green on the right, were just the opposite, meaning the two aircraft were on a collision course. He immediately leveled the wings and pulled back hard on the yoke, flying over the other Skywarrior, barely avoiding a mid-air catastrophe. On the next attempt he was able to refuel the lost aircraft and accompany it back to the carrier.

In spring 1975, two years after flying his last combat mission of the conflict, Mike was a lieutenant stationed at NAS Agana in Guam with VQ-1, flying the EA-3B. It was Saturday 26 April 1975 and he had just sat down and popped open a beer when the CO of the squadron, Captain Tim Connolly, hastily appeared and told him "You are opening a refugee camp tonight, and 550 people will arrive in five hours. You need to go talk to XO Commander Bolt right now!"

It was appropriately called *Operation New Life*, the evacuation of more than 100,000 South Vietnamese refugees to Guam for processing before beginning their new lives in the United States and Canada. The Navy flag officer in charge of this operation, Rear Admiral George Morrison, was Father of The Doors frontman Jim Morrison. The younger Morrison had famously passed away in Paris on 3 July 1971.

After processing the first group of refugees, a second group arrived at midnight two days later. Upon arrival, Mike asked if there was somebody amongst them who spoke English and could act as a liaison. One seemingly nondescript refugee stood up and volunteered. His name was Nguyen Cao Ky, past commander of the VNAF and former Prime Minister and Vice President of the Republic of [South] Vietnam. Mike recalls that over the following days Ky worked tirelessly to reunite families that had been separated during the chaotic evacuation. Ky himself eventually settled in Southern California.

EKA-3B pilot Lieutenant Scottie Atkins also flew out of Da Nang during his 1969-1970 deployment as part of VAQ-135, with the squadron formally attached to CVW-15 on *Coral Sea*. Amongst his memories of Da Nang is the tropical heat and humidity, which at times could be stifling:

> Due to congestion at the airfield, the time between starting our engines and taking off could be as long as 45 minutes. As the air conditioning packs in the A-3 were run by engine bleed air, sitting on the ground at idle produced barely a whisper of cold air. In hot, humid weather so much sweat would drip from my hands that a puddle would form on the deck in front of my seat. I would have to put a towel down there to soak it all up.
>
> The standard-issue sidearm for us was a .38 caliber revolver that was usually worn in a shoulder holster. We often joked that it was such a small caliber the only thing

it was good for was shooting ourselves if we were about to be captured! We were supposed to take care of it, with regular cleaning and oiling. I didn't, and by the end of the deployment when I took it out of the holster, it was completely corroded from my perspiration. It is a good thing I never had to rely on it to defend myself as I doubt it would have fired.

Atkins recalls a night departure from Da Nang that was particularly memorable for him, and his crew:

It was SOP that when taking off at night we would turn on all the aircraft lights. One night we took off from Da Nang, and as we climbed out we saw multiple tracers coming from the ground. I called the tower and told them somebody was shooting at us. They replied it wasn't enemy fire, it was some U.S. Marines *playing with us*. The next night we took off again, but unlike the previous night, I didn't turn the lights on. Predictably the tower called and told us our lights weren't on. I replied, "There's a reason for that!"

Atkins was flying the duty tanker on 16 May 1970 on a routine mission to refuel **BARCAP** and **FORCECAP** fighters protecting CTF-77. While 40 miles northeast of Da Nang and setting up to land, he found himself directly behind the aircraft being flown by Skeen, McNally, and Conner:

There were thunderstorms in the gulf that day, which at those latitudes can reach high into the stratosphere. The lower third is where they are most intense, with heavy downpours as well as powerful microbursts and downdrafts. My NFO Lieutenant Patrick "Pat" Staples pulled the hood off his radar scope and showed me the storm. It was the largest mass I had ever seen. The storm was concentrated over the water, and did not extend far inland, with Da Nang reporting VFR conditions. We were on a westerly heading at 20,000 feet, and made a left turn to a southerly heading to avoid the worst of the storm, and set up for landing to the north at Da Nang. The XO, who was ahead of us, continued straight into the storm. This baffled me, as his navigator McNally was an experienced and competent NFO, and would have seen the same picture on his radar we did. However, due to costs and parts availability, the radars weren't always working. This left a chance his was not operating that particular flight, and thus may not have been aware of the severity of the storm.

When we came into the break for landing in Da Nang, I looked down and saw an A-3 in the hot fueling pits. I thought to myself that must be the XO, as he liked to keep the speed up when he flew, and perhaps he landed to the south and arrived before us. After landing and shutting down, I asked if that was the XO's airplane in the pits, and the answer was no. Hue, which had an American military airfield, is about 50 miles northwest of Da Nang. Somebody made a phone call and inquired if an A-3 had landed up there, and the reply was no. Other airfields in the area were also contacted and none of them had heard from, or seen the aircraft. At that time *Coral Sea* and other boats in the gulf were contacted, along with SAR forces, and a rescue effort was launched. My crew and I also launched that night to help in the search. The destroyer *Waddell* eventually found aircraft wreckage in the water, along with McNally's remains.

After graduating college Captain Terris "Terry" Hanson entered AOCS in Pensacola. Following primary flight training, he continued in the pipeline to Meridian, Beeville and Miramar, before arriving to RAG VAH-123 at Whidbey Island, and subsequently VAQ-130 at NAS Alameda. He first deployed to Vietnam in May 1971 with Det. 2 as part of CVW-5 on *Midway*, piloting an EKA-3B.

*Midway* had completed one Vietnam combat cruise spanning March-November 1965. During that deployment the carrier launched combat sorties from both Dixie and Yankee Stations against targets in South and North Vietnam. On that cruise, which coincided with the opening salvos of *Operation Rolling Thunder*, *Midway* lost 22 aircraft, resulting in the deaths of 12 airmen. Five additional airmen were captured as POWs. Following that deployment, the carrier entered Hunters Point Naval Shipyard in San Francisco for an extensive modernization program. After four years the process was nearly completed, however upgrades to the WWII-era carrier unintentionally created many new problems. Terry remembers:

The modernization included installing the largest flight deck area of any supercarrier, including the widest angle and crosswind for landing aircraft—13 degrees versus 11 degrees. This large flight deck stacked atop a medium-size hull caused *Midway* to do unusual gyrations, such as the fantail doing a constant horizontal figure-eight which, needless to say, created challenges for landing aircraft. Adding to the problems, the funds to finish the project ran out before completion. Therefore, the island was not moved aft to longitudinally counterbalance the extreme weight of the flight deck angled overhang on the port beam. To counter this, a great amount of concrete ballast was poured starboard and aft to correct for stability, yet furthering the unusual gyrations of the hull and flight deck.

It has been said that engineers calculated that with this top-heavy deck and weight of the angled overhang, in heavy seas if the boat rolled port beyond a certain point, it would continue to roll and capsize. To prevent that possibility, explosive bolts were attached to the angle overhang of the flight deck, and the skipper had a button on the bridge to jettison that portion if the carrier exceeded a certain degree of roll. Putting it mildly, landing aboard *Midway* provided for lots of excitement.

Terry's first combat cruise aboard *Midway* in 1971 saw the air wing lose two aircraft, neither to combat causes. The first was an E-2B Hawkeye (BuNo. 151719) on a ferry flight, with all five souls aboard lost. This was followed by a KA-6D Intruder (BuNo. 152598) that had to be abandoned in-flight due to a fire in the tanker package. Both crewmen were rescued.

*Midway* returned to Alameda in autumn 1971 and began preparation for a third Vietnam deployment, originally scheduled to begin the following June. On 30 March 1972 the carrier, along with CVW-5, were performing ORE/ORI off the coast of California when North Vietnam launched the Easter Offensive. The exercises were abruptly cancelled, and the carrier ordered back to NAS Alameda. CNO Admiral Elmo Zumwalt flew in from Washington, DC aboard a TA-3B VIP transport and, upon arrival, came aboard *Midway* to give a short, inspirational speech to the assembled crew. The carrier departed Alameda on 10 April full-steam for Yankee Station, making an unscheduled stop in Pearl Harbor to repair sponsons ripped off by a storm encountered in transit.

Arriving on Yankee Station, operations to stem the tide of the invasion were frenzied efforts not seen since the Tet Offensive. Rather than the normal seven launch cycles per day/night involving about a dozen aircraft per cycle, the air wing was launching no less than three Alpha strikes per day/night cycle, with extended mission times. Each Alpha strike consisted of about 36 aircraft, launched and subsequently assembled in-flight. In addition to refueling dozens of aircraft in these massive strikes, EKA-3Bs were also jamming North Vietnamese radars and radio communications. Adding to the difficulty and complexity for *Midway* and the air wing was the fact that the carrier had only two catapults yet were expected to, and did keep pace with, larger supercarriers equipped with four catapults.

Terry remembers tanker sorties accounting for about 75% of missions flown that second cruise on *Midway*, with jamming accounting for the remainder. When jamming missions were flown it was always in conjunction with tanker duties. Despite the EW capabilities of the EKA-3Bs proving their worth time and time again, a modest number of fighter pilots were opposed to their use, feeling that "All the jammers do is alert the enemy that we are coming". He has many memorable moments from his combat deployments:

> The most memorable missions for me would be the occasions of dodging SAMs and AAA, particularly at night when visibility and depth perception were limited. The PAVN commonly launched SA-2s in volleys of 2-3, fired 1-2 seconds apart, as the system could track up to three missiles on a single target. We would wait for the first SAM to get close, then break hard at the last second as the missile couldn't turn as tight as the aircraft. After dodging the first SAM we knew that more were likely coming, which was rather disconcerting. Dodging SAMs in a Skywarrior, at night over water, definitely got the adrenaline pumping.

The 1972-1973 *Midway* WESTPAC covered the last two major air offensives of the conflict, *Linebacker* and *Linebacker II*. As a result the carrier lost 15 aircraft to combat causes, and 5 to operational incidents. These losses resulted in the deaths of 10 airmen, with a further 6 taken as POWs. Terry recalls a tragic incident that cruise that was particularly painful for *Midway* and CVW-5.

An A-6 that had just refueled over the Tonkin Gulf. *Coral Sea* is in background.

On the night of 24 October 1972 an A-6 (BuNo. 155705) of VA-115 *Arabs* was returning from a night bombing mission over North Vietnam. Four Intruders had taken part in the strike, two as bombers and two as Iron Hand SAM suppression. The aircraft was being flown by pilot Lieutenant Bruce Kallsen and B/N Lieutenant (junior grade) Michael Bixel. While Kallsen was on his third combat deployment to SEA Bixel was on his first. Both men were scheduled to go on leave in the following days. In addition to being a dark, moonless night, heavy swells resulted in the boat heaving in the seas, and a stiff crosswind blew across the flight deck. An additional complexity were the two Mk.82 bombs that would not deploy, and were hung on the furthest outboard starboard wing station, putting the aircraft at the asymmetric load limit.

Normally, an aircraft with hung ordnance would bingo to Da Nang, or the entire MER would be jettisoned at sea. However, due to numerous factors, the men were directed to come aboard, despite the risk of trapping with unexploded ordnance. This would prove to be tragically consequential.

A hard landing resulted in the starboard axle shearing off.[81] The Intruder careened out of control down the flight deck, veering starboard towards parked aircraft and personnel. Air wing CAG Commander Myers had climbed down from the F-4 (BuNo. 153031) he had just landed when he was struck by the Intruder, severely injuring, and nearly severing his leg. As the aircraft continued its path of destruction, it became evident to B/N Bixel he would be seriously wounded or killed in the inevitable collision with a row of parked A-7s. He ejected, despite being outside the envelope to safely do so.

The A-6 finally came to a halt after hitting the Corsairs, its wing cutting through an A-7 canopy and scraping the helmet of the pilot in the cockpit who had just landed and shut down the engine. In total five men lost their lives in the incident, four of whom were flight deck personnel. Bixel, who after ejecting landed in the water, was never located. Kallsen miraculously escaped without serious injury.

*Midway* set the record for the most days on the line by a carrier in the conflict that deployment, an astounding 208. Their extraordinary performance and devotion to duty resulted in the boat and crew being awarded the prestigious Presidential Unit Citation, a unit award equivalent to the individual Navy Cross. Terry credits this achievement to the exceptional leadership of Skipper Captain Sylvester Foley, who would advance to admiral and later become COMPACFLT. Of Captain Foley, he remarks:

A most superior leader, he put particular confidence and trust in certain EKA-3B pilots. It was a very difficult cruise for the men aboard *Midway*—to support the entire air wing in missions over North Vietnam, and then necessarily and expeditiously get them back aboard the carrier, under the most challenging of circumstances, in all weather conditions, sea states, day and night. His leadership of *Midway* included the record for the most consecutive days on the line in North Vietnam by a carrier, which resulted in the boat being one of just a few carriers to receive the Presidential Unit Citation. This was done with only two catapults, unlike most other big carriers with four catapults. *Midway* sustained the equivalent or greater mission aircraft launches/sortie cycle of all those other big deck carriers that had simultaneous launch capability.

*Midway* would gain worldwide notoriety in April 1975 during *Operation Frequent Wind*—the chaotic evacuation of Saigon. The surreal images of helicopters being pushed off its flight deck into the ocean to make way for VNAF Major Buang-Ly, his wife, and five children, to land aboard are some of the most iconic of the 20th century, and cemented the carrier's place in American history. The O-1 Birddog aircraft flown by Major Buang-Ly is currently on display at the National Naval Aviation Museum in Pensacola, Florida.

After graduating college, Captain Chuck Hanson signed up with the Navy as an NFO. Following six weeks in Pensacola, he was offered the opportunity to attend pilot training. After earning his wings, he arrived to VAQ-130 Det. 4 as a nugget, flying the EKA-3B, part of CVW-14 aboard *Enterprise* for the carrier's June 1971-February 1972 WESTPAC cruise. As that cruise concluded, CVW-15 aboard *Coral Sea*, which had recently begun a December 1971-July 1972 WESTPAC deployment, found themselves short two pilots in VAQ-135 Det. 3. This was a result of one pilot unexpectedly retiring, while the other was found to no longer possess the necessary skills to trap aboard the carrier at night. An urgent request was put out for volunteers to augment the squadron, with Chuck and Lieutenant Craig Porterfield answering the call. Both men disembarked from *Enterprise* in Subic Bay, and a few days later boarded *Coral Sea*, which would soon be returning to Yankee Station. It was during this second deployment that Chuck experienced an unforgettable event that is forever etched into his consciousness.

It was 23 April 1972, three weeks after North Vietnam launched the Easter Offensive. On that day Chuck returned to *Coral Sea* after ferrying an EKA-3B from Cubi Point that had undergone routine maintenance. After landing on the carrier late that afternoon, detachment OinC Lieutenant Commander Harvey "Harv" Dickey met him on the flight deck, telling him to get some sleep as he was scheduled for a mission early the following morning, with a briefing to begin at midnight. The mission, part of *Operation Freedom Train*, was an ECM sortie in support of B-52[82] *Arc Light* air strikes on port facilities at Thanh Hoa. A total of nine EKA-3Bs would take part in the mission, three each from *Coral Sea*, *Hancock* (VAQ-135 Det. 5) and *Constellation* (VAQ-130 Det. 1). *Hancock's* three EKA-3Bs took off from FASU Da Nang as they were not billeted aboard the *Essex*-class carrier.

As Chuck did not have the appropriate security clearance, he could only attend the second portion of the briefing, covering the flight. The first portion, covering EW, was given only to NFOs who would be operating the sophisticated jamming equipment. Chuck's crew that night were two men he did not normally fly with—Lieutenant Doughlas Kees and Lieutenant (junior grade) Steve Kuhar. Kees would be in the right seat operating the ALQ-92 radio jammer, while Kuhar would occupy the third seat operating the ALT-27 radar jammer. Around midnight Steve snapped a candid picture of Chuck in the ready room they shared with helicopter squadron HC-1 Det. 6. The photo clearly shows a very tired and worn man.

Their orders were to take up station feet wet off the coast of Thanh Hoa and fly a racetrack holding pattern, performing standoff radar and radio jamming. As this mission was in support of the Air Force, Det. 3 Whales taking part changed their call sign from "Lightening Bolt" to "Nagel". At 0200 on the morning of 24 April Chuck, Doug Kees and Steve Kuhar, along with two other Det. 3 Skywarriors, catapulted from *Coral Sea*. While Chuck and Doug were on their second combat deployments, Steve was on his first. He had originally joined the Navy straight out of college, with the intention of becoming a pilot. However, the stringent vision requirements eliminated that possibility. Arriving to Pensacola on St. Patricks Day 1970 he was intent on becoming an A-6 B/N. However when graduation came there was only one coveted Intruder B/N slot available, which was taken by the first in class.

Though operating the third-seat radar jammer on this mission, Steve also routinely flew in the right seat, performing co-pilot duties as well as radio jamming. He relates that there were times when he would be monitoring VPAF radio traffic and hear surprising transmissions:

Chuck Hanson in ready room aboard *Coral Sea* in the early morning hours of 24 April 1972.

Steve Kuhar next to a Whale on *Coral Sea*.
This picture shows the immense size of the A-3.

There were times I would be listening in on VPAF radio traffic, and be able to vaguely make out what they were talking about, as they did not translate squadron call signs. I would hear them say "Old Nick" and would know they were talking about F-4s of VF-111. It was odd and rather disconcerting they used squadron call signs, as if they knew exactly who was coming.

Once airborne, the flight of Skywarriors climbed to 5,000 feet and obligatorily checked in with Red Crown on *Long Beach* for vectors to station. *Long Beach* was the nuclear-powered guided-missile cruiser picket ship in the Tonkin Gulf performing, amongst other duties, air traffic control services for combat aircraft over the Tonkin Gulf and North Vietnam. As their Skywarrior (BuNo. 142403) did not have a functioning radar, they were wholly dependent on *Long Beach* to vector them to the right location for the mission. This would ensure the aircraft stayed feet wet and out of range of AAA and SAMs.

Red Crown instructed Chuck to climb to 17,000 feet, and fly a heading of 310° and distance of 185 nautical miles to take up station twenty miles off the coast. As an EMCON "ziplip" condition had been placed on the mission, they began observing radio silence after receiving their vectors from Red Crown. This resulted in the normally active communication channels being eerily quiet. The weather that moonless night saw a solid undercast layer at 12,000-15,000 feet, with clear skies above. Therefore the Skywarriors could not see the ground below them, only a blanket layer of clouds. Once reaching station at 0230, Chuck began flying a northwest-southeast racetrack pattern, while NFOs Doug and Steve went to work jamming radars and VPAF radio frequencies.

To maximize effectiveness of the jammers, Chuck flew the Skywarrior with half-flaps and airspeed of 170 knots.

Doug Kees was the senior officer onboard, and a highly experienced navigator in the A-3. He was simultaneously plotting their course on his kneeboard, using dead reckoning in an attempt to keep track of their location the best he could without radar, daylight, or visual landmarks. Steve remembers Doug voicing his concern with the course and distance they had been given by Red Crown. Something just didn't feel right to his well-honed navigation skills.

While flying a southeast leg of the holding pattern, and believing they were safely over the Tonkin Gulf, the amber light on Steve's RHAW console began blinking, indicating they were being painted by Spoon Rest acquisition radar. The light was accompanied by slow warble tones in their headsets. On the guard channel he routinely transmitted "SAM! SAM! Vicinity of Thanh Hoa. Nagel 6 out!"

Within seconds of making that first transmission, the red RHAW light began blinking, accompanied by fast warble tones, meaning missiles were in the air and Fan Song guidance radar was guiding them to their targets. He made a second transmission "SAM launch! SAM launch! Vicinity of Thanh Hoa. Nagel 6 out!" Steve remembers:

> SAM warnings on the guard channel were standard fare, and while flying over the Tonkin Gulf we would not normally be concerned with being in danger from SA-2s, as we were typically offshore. However, Chuck said something that made me look out the window, and down to the undercast. When I did I saw a huge flame ripping through the clouds and a missile screaming right for us.

Chuck recalls what happened:

> I casually looked down and saw a missile shoot out from the undercast cloud layer at our ten o'clock position. Believing we were over the Tonkin Gulf I remember remarking "Oh look, our ships are sending in missiles now". In my peripheral vision I saw Steve turn and look out the window. He ripped off his oxygen mask and yelled "SAM! SAM low!"
>
> My next action, instinctive from training, was to break into the missile. I turned the yoke hard to port and pulled into the SAM's trajectory, turning about 130°. Because I was flying slow with half-flaps, the high bank angle caused the nose of the airplane to drop below the horizon, and we began descending toward the undercast with increasing speed.
>
> I saw a flash as the missile passed by us in the night sky, off our starboard side and about 100 yards away. It looked exactly how it has been described—a flying telephone pole with fins and a flame shooting out the back that was as long as the missile itself. I retracted the flaps to prevent overspeed damage, pulled back the throttles to idle, and deployed the speed brakes in a bid to try and slow down the airplane and correct our attitude.

As I was struggling with the airplane, I remember a particular transmission coming over the guard channel, which by then was a flurry of activity. It was an EA-3B of VQ-1 broadcasting "This is Deep Sea on guard relaying for Motel. SAM! SAM! Vicinity of Thanh Hoa. Deep Sea out!"

Despite idling the engines and deploying the speed brakes, our airspeed was still increasing. We were just about to descend into the undercast when a second missile broke through. It screamed by our starboard side so closely that the rocket engine bathed the cockpit in bright light. I heard a roar similar to an afterburner, and felt a vibration as it shot past us. I remember thinking that if it had detonated at that distance it surely would have blown us to smithereens.

I maintained focus on correcting the attitude of the airplane, as we were still losing altitude, hurtling toward the ground. We had just descended into the cloud layer, and were in instrument conditions, when a third missile shot past. Due to having no visibility out of the cockpit, I didn't see it and don't know exactly how close it came, but it was close enough to light up the clouds around us.

With no visual horizon, I began concentrating my instrument scan on the gyroscopic attitude indicator, which was vital to recovering the aircraft. I recall it showing all black, meaning the Whale was descending nearly vertically. Other instruments such as airspeed, vertical speed, and altitude, were either pegged-out or winding wildly.

While Chuck was struggling to recover the aircraft, Steve was hanging on for dear life in the third seat:

> Facing rear and looking up, I could feel the airplane rolling, and us descending quickly. It felt like we were literally falling out of the sky. When we finally broke through the bottom of the overcast, I looked out the window and was shocked to see that instead of being over the Tonkin Gulf as we should have been, we were instead over Thanh Hoa! Then I saw them, what looked like tennis balls flying around us. Holy hell! That's AAA! By that time I had put my oxygen mask back on, and was breathing so heavily I thought I was going to suck the rubber right off it.

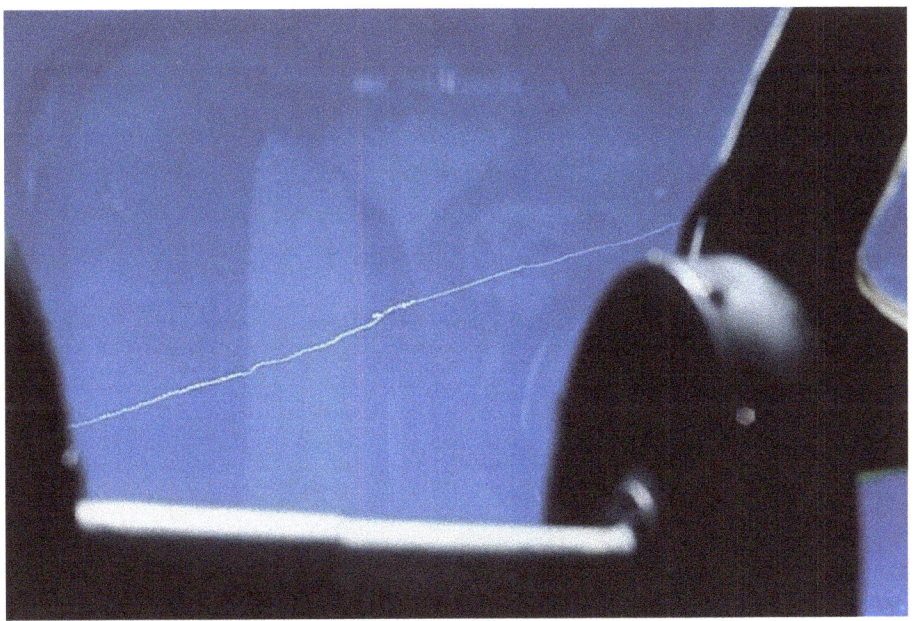

SA-2 SAM contrail as seen from A-3 cockpit.

Chuck had no time to react to the anti-aircraft fire as he was concentrating with all his might on recovering the aircraft before it, and those aboard, became additional casualties of the waning conflict:

> We had just broken through the bottom of the overcast when I managed to roll the wings level and began pulling back on the yoke to recover from the dive, while retracting the speed brakes and adding full power. I eventually saw a sliver of white at the top of the attitude gyro, which grew. I continued to pitch up, the nose reached the horizon, and we began to gain altitude as we climbed back into the overcast.
>
> After breaking out on top of the clouds we climbed past our original altitude of 17,000 feet. It seemed like we had escaped the dangers when I noticed multiple black objects falling from above. As we continued to gain altitude I looked up into the sky for the source. Some distance above, I spotted the red anti-collision lights of two F-4s. They were tucked snugly under the wings of a B-52 BUFF, it's bomb bay open and empty. Those falling black objects I had seen were iron bombs being dropped by the Stratofortress. We had just flown through its bomb stream! I stopped climbing, turned the aircraft due east towards the gulf, and got us the hell out of there.

Steve remembers that as they were egressing from the area, Doug voiced relief that his suspicion of the course and distance given to them by Red Crown had been validated, excitedly repeating several times "I knew we were in the wrong spot!"

On that morning, *Coral Sea* was operating with only three of its four propeller shafts. As the A-3 typically required a minimum of 26 knots of headwind down the flight deck for launch and recovery, this left the boat unable to generate sufficient wind. To remedy this, the carrier turned into a squall line of thunderstorms to take advantage of the downdrafts created. Steve was still reeling from their near-death experience when he looked out the window and saw ominous dark clouds, and choppy seas as the aircraft approached the carrier:

> We had dodged three SAMs, a death dive towards the ground, AAA, and the bomb stream of a B-52. After all that, we were flying straight into a thunderstorm! I was naturally worried, but Chuck is an excellent pilot and he landed on the first attempt, leaving the three of us breathing a tremendous sigh of relief. My roommate on that cruise was Mike Danosky, the pilot I normally flew with. When I got to our stateroom early that morning, he was sound asleep. My adrenaline was still pumping, and I was eager to tell somebody about all I had just been through. Mike's slumbered response was to tell me to shut up and talk to him about it later!

To Chuck, the squall line was one more hurdle to clear on a night filled with surprises:

> I was operating with little sleep, the trauma of nearly dying, and a burning desire to exit the cockpit. Despite a severely pitching deck, I managed to grab the No. 1 wire, garnering a "cut pass" grade from the friendly LSO. We had just shutdown the whining J-57 engines when I heard the 1MC penetrate loudly through our open overhead hatch, "Lieutenant Hanson report to the bridge, *immediately!*" From what I remember, Doug Kees took the brunt of the questions, as he was the senior officer onboard.

An investigation was opened into just how the EKA-3B had ended up some ten miles feet dry over Thanh Hoa, when they were supposed to have been flying twenty miles off the coast feet wet. The fault was placed on Red Crown, who had not only given Chuck incorrect navigation to station, but had also failed to warn the crew they were flying over one of the most heavily defended targets in the world-Dragon's Jaw Bridge at Thanh Hoa. It was later discovered that the controller who had given Chuck course and distance was inexperienced, and should not have been working unsupervised. Due in part to the unconventional call signs assigned by the Air Force, the man had mistakingly given the EKA-3Bs vectors meant for F-4 fighters performing MiGCAP. The investigation resulted in Non-Judicial Punishment (NJP) being quietly meted out, and the incident subsequently and quickly swept under the rug.

As to why none of the three missiles detonated, the men have several theories. As numerous ECM aircraft were in the air, one or more could have jammed the operator's radar. Another possibility is that the missiles did not have time to arm before reaching close proximity to their target. It is also possible that the Skywarrior, a large target flying slowly directly over Thanh Hoa, confused the operators. A less factual but more likely theory is that in the grand scheme of the universe, it simply was not their time.

For Chuck and Steve, the events over Thanh Hoa that morning which nearly cost them their lives seemed so surreal, and incomprehensible, that for many years after they questioned if they had taken place as they remembered. Several decades later, the two men ran into each other at an A-3 reunion in San Diego, the first time they had spoken since serving together on *Coral Sea*. Chuck's first words to Steve were "Do you remember the night over Thanh Hoa?" Steve recalls his answer:

> I told him yes I remembered, however before discussing I asked him to tell me his recollections of the events. What he described matched my memories, validating for both of us what happened that dark night so long ago. Subsequently, I showed my children and grandchildren the picture I had taken of Chuck in the ready room, just before we left for that mission. I told them if not for this man, I would not be here today. He saved my life, and made yours possible.

On 4 April, just three weeks prior to the incident over Thanh Hoa, Steve and Mike Danosky had been flying an ECM mission. That particular mission was unusual for several reasons. They were feet dry directly over the DMZ, which was an area of intense combat following the Easter Offensive. It was also much further south from where they usually flew their jamming sorties. Lastly, they were flying an east-west jamming track instead of the normal north-south tracks they had flown further north.

While over the DMZ, they heard an Air Force aircraft with the call sign of "Bat 21" check in with the controlling agency. Shortly after hearing this transmission, the men completed their mission and received radar vectors back to *Coral Sea*. After crossing the coastline feet wet, they heard the distinctive siren-like sounds of locater beacons on the guard channel. The beacons were coming from the area of Cam Lo, fifteen miles inland from the coast, and just south of the DMZ. Bat 21, which was a USAF EB-66C Destroyer (Serial No. 54-0466) from the 338th TFW/42nd TEWS with six men aboard, had been brought down by a SAM. The only survivor from the aircraft, navigator Lieutenant Colonel Iceal "Gene" Hambleton, ejected and landed just south of the DMZ, amongst an estimated force of 30,000 NVA soldiers invading South Vietnam. Becoming the longest SAR mission of the conflict, for five days the U.S. military tried to extract Hambleton by air, which was unsuccessful and cost eleven U.S. airmen their lives. Finally, an audacious ground rescue was planned, led by Navy SEAL Lieutenant Thomas Norris, who was accompanied by VNN Petty Officer Nguyen Van Kiet. After eleven days both Hambleton and USAF 1st Lieutenant Mark Clark,[83] the crewman of an OV-10 Bronco (Serial No. 68-3789) that was shot down during the SAR operation, were rescued. Steve and

Mike flew multiple jamming sorties over those days in support of the operation. Several weeks after the incident, the Air Force acknowledged the value of the Navy's assistance in the rescues. Norris was awarded the Congressional Medal of Honor, while Van Kiet received the Navy Cross.

The urgency and costly sacrifices to extract Hambleton were due to his intimate knowledge of American ICBM technology. He had previously worked on the PGM-19 Jupiter, Titan I and Titan II ICBMs. He had also commanded the 571st Strategic Missile Squadron at Davis-Monthan Air Force Base in Arizona 1965-1971. Had he been captured, he would have likely been turned over to the Soviets, never to be seen or heard from again.

Steve recalls another incident that cruise involving Chuck and Doug Kees. One day, he and pilot Lieutenant Bill Lee were flying a mission over the Tonkin Gulf, when they found themselves dangerously low on fuel. As a result, they declared an emergency and took a heading to Da Nang. Steve remembers:

> I was doing constant fuel calculations on my kneeboard, and no matter which way I ran the numbers, we were not going to make it to Da Nang or anywhere else. I began tightening my harness and eyeballing the handles to blow the belly hatches. I also began rehearsing the egress procedure in my head, for if we ditched in the water time would be of the essence.

As they neared the coast, the men spotted the parallel north-south runways, giving a glimmer of hope that if they egressed over the water they were close enough to Da Nang for a quick rescue. Then, like a scene out of a movie, Steve looked out the starboard window and saw Chuck and Doug come from out of the wild blue yonder. They skillfully maneuvered in front of the thirsty A-3 to begin delivering life-saving JP-5, resulting in Steve and Bill landing safely at Da Nang. "That is another example of Chuck Hanson's actions resulting in the sparing of my life" Steve gratefully remarks.

Along with the night over Thanh Hoa, the rescue of Hambleton, and nearly running out of fuel en route to Da Nang, there was another mission that April that stands out in Steve's mind for entirely different reasons. One day that month, he was flying with Mike Danosky on a tanker sortie over the Tonkin Gulf. On their return flight Steve routinely checked in with *Coral Sea*, and after an exchange of formalities, the carrier oddly inquired as to if the men had any cigars onboard. Sharing a puzzled look with Mike, Steve asked for the question to be repeated, and again they were asked the same question. Believing it was a security challenge, Steve began frantically looking through his paperwork for the appropriate response. However, there was no mention of cigars, or a challenge question. After receiving the query again Steve sheepishly answered no, there were no cigars onboard. The voice came back, impatiently asking for confirmation that Lieutenant Danosky was flying the aircraft. When Steve confirmed Mike was indeed the pilot, the controller on *Coral Sea* offered congratulations, informing them that Mike's wife had just given birth to a healthy baby boy in Alameda. Both men breathed a joyous sigh of relief.

At 2100 hours EDT on the evening of 8 May 1972, President Nixon addressed the nation in a televised speech. As he spoke, there were numerous noticeable pauses, reminiscent of President Johnson's televised announcement of *Operation Pierce Arrow* attacks on 4 August 1964. Nixon was carefully timing his speech to coincide with *Operation*

*Pocket Money*, a maritime mining operation, years in the making, targeting the ports of Haiphong, Dong Khoi, Quang Khe, Thanh Hoa, and Phuc Loi. As Nixon began his address, in the Tonkin Gulf Steve Kuhar and Mike Danosky catapulted off *Coral Sea*. They were the lead aircraft of a group from the carrier that included three A-6A Intruders of USMC attack squadron VMFA(AW)-224 *Fighting Bengals*, each carrying four 1,000-pound Mk.52 magnetic mines. In addition, there were six A-7s, three each from VA-22 *Fighting Redcocks* and VA-94. There were also several F-4s providing MiGCAP. Steve recalls the operation:

> Navy Commander Roger "Blinky" Sheets was flying the lead Intruder. Sheets was very studious and meticulous, while his B/N Captain Charlie "Vulture" Carr was a swaggering and hard-nosed Marine with a wild hair; a larger-than-life character who relished flying combat. I remember one day on *Coral Sea* when Charlie excitedly called me over to his A-6 after completing a combat mission. The canopy had holes in it, and there was mud in the cockpit. He dropped the bombs so low he fragged his own aircraft. He was quite proud!
>
> The mining operation involved flying at wave-top level into Haiphong Harbor to avoid radar and early detection. We were lucky, because from a meteorological standpoint, the weather was perfect for that tactic. As we approached the harbor, the F-4s began accelerating so they could quickly climb in altitude to perform MiGCAP. We also began to accelerate in anticipation of climbing in altitude to perform jamming.

> Being that the Phantoms were much faster than the Whale, they pulled ahead of us. I distinctly recall looking out the starboard window at a pair of F-4s who were flying past us in formation. We were flying so low that as they went by I saw their rooster tails in the water. That picture is still very vivid in my memories.

The mining operation that day was a resounding success, with the only substantive resistance apart from AAA being two VPAF MiGs that were vectored to intercept the minelayers. One of those MiGs was shot down by USS *Chicago* (CA-136) using a RIM-8 Talos missile, sending the second MiG fleeing. The operation was not, however, free from tragedy. The evening prior, Rear Admiral Rembrandt Robinson, his chief of staff Captain Edmund Taylor, Jr. and operations officer Commander John M. Leaver, Jr. were killed in a helicopter crash at 2245 hours. They were returning to USS *Providence* (CLG-6) from a mission planning meeting aboard *Coral Sea*. Admiral Robinson was the eighth, and final flag officer fatality of the conflict.

Steve's second cruise on *Coral Sea* with VAQ-135 Det. 3. was after the 27 January 1973 ceasefire. However, despite the cessation of hostilities with North Vietnam, there was still an important task to be accomplished. Over the course of the conflict the Navy and Marines dropped over 11,000 maritime mines in North Vietnamese harbors, coastal waters, and rivers. The removal of those mines was accomplished February-July 1973 during *Operation End Sweep*. He flew numerous missions during the unprecedented operation, performing stand-off jamming while helicopters and surface ships swept for mines. Although the United States and North Vietnam were cooperative in the removal of the mines, Steve shares that during the operation American aircraft were constantly painted by North Vietnamese SAM radars. Perhaps weary of reigniting the wrath of President Nixon, and a possible dreaded return of B-52s over the country, not a single munition was fired at *End Sweep* aircraft.

As with many who served in the Vietnam theatre, Steve experienced several painful losses. There is one in particular that he continues to think of often. On 6 April 1972 Commander Thomas Dunlop, CAG of CVW-15, had been flying a road recce mission in an A-7 (BuNo. 157590) call sign "Beefeater 300" of VA-22 from *Coral Sea*. Dunlop and his wingman were about seven miles north of Dong Hoi when they encountered intense anti-aircraft defenses. His Corsair took a direct hit from an SA-2, destroying it. No parachute was observed and no beacon or voice communications were detected. It is likely that Dunlop was killed instantly. He was the sixth, and final, CAG killed during the conflict. After his tragic loss, Commander Sheets assumed the role of CAG.

Steve remembers being on the flight deck that fateful Thursday, preflighting for a mission, when Dunlop stopped by, warning him to be careful. Both men took to the air soon after, with Steve performing radar jamming, protecting Dunlop and others from SAM threats. Despite the countless lives he saved from the deadly missiles, he continues to question whether there was something else he could have done that day to avoid Dunlop's loss. In the many decades since, whenever he has been in the nation's capitol, Steve has made a point to visit the Vietnam Veterans Memorial Wall, where he pays tribute to Commander Dunlop, and others lost.

It was on the 1973 cruise, the seventh and last Vietnam WESTPAC for *Coral Sea*, that Steve and Mike Danosky found a spark of solace amidst the chaotic American withdrawal. As *Operation Homecoming* was underway, the two ferried four F-4 crewmen from *Coral Sea* to Clark Field in the Philippines. There they were able to welcome their shipmates back from the hell of Hanoi Hilton, and other North Vietnamese POW camps:

> Those four men were beyond ecstatic. As soon as the A-3 came to a stop, they jumped out of the aircraft and literally ran to welcome back their friends, many of whom they thought they would never see again. That was definitely one of the happier, and more gratifying, moments of my naval service in SEA.

For the United States, there were very few celebratory occasions during the conflict, and *Operation Homecoming* was one of them. However, the joyous return of 591 POWs from North Vietnam was tempered by the painful losses of more than 58,000 American service men and women who gave their lives. This is in addition to the thousands still classified as missing at the time the peace accords were signed. Tragically, more than half a century later, the fates of many of those missing remain unknown.

The Skywarrior continued to serve faithfully in the decades following the Vietnam conflict until fall 1988, when planning began to "sunset" the aircraft. The Whale's last combat deployments occurred during the 1990-1991 Gulf War. By that time there were nine variants of the aircraft, and all were officially retired from the Navy in September 1991 after more than 35 years of dedicated service. Several static A-3 airframes are on display around the country. The sole surviving EKA-3B is on the floating *Midway* museum in an ancestral home of Naval Aviation, San Diego.

Despite a contentious safety record, and a number of operational incidents resulting in loss of life, the men who flew A-3s in combat resolutely stand by the aircraft. Those who served in the Skywarrior, along with the countless lives saved by its presence, consider it one of the most vital, yet under-appreciated aircraft to ever serve in the fleet.

---

69 Heinemann, Ed. *Ed Heinemann:Combat Aircraft Designer.* United States Naval Institute 1980.

70 NASA astronauts Roger Chaffee, Ed White and Virgil Grissom died on 27 January 1967, the result of a capsule fire during an Apollo launch pad test at Cape Canaveral. On 24 November 1969 Ed White's younger brother USAF Major James White of the 355th TFW/357th TFS based at Takhli RTAFB was shot down over Laos. Flying an F-105D Thunderchief (Serial No. 61-0060), he was part of a two-aircraft formation on an *Operation Barrel Roll* mission when his aircraft disappeared. His remains were recovered between 2010 and 2016 and interred at West Point Cemetery in June 2018.

71 The Navy standardized on JP-5, which was less volatile and had a higher flash point temperature than JP-4 used by the Air Force. JP-5 is also slightly heavier at 6.8 pounds per gallon versus 6.5 pounds per gallon for JP-4.

72 Following WWII the Navy began a process to modify *Essex*-class carriers to accommodate jet aircraft, which were larger, heavier and faster than their piston counterparts. There were two modification programs—SCB-27A and SCB-27C. The most significant difference between the two was hydraulic versus steam catapults.

73 Narrative of events from letter by Captain Gerard Colleran, Commanding Officer *Bon Homme Richard* to Commander-in-Chief U.S. Pacific Fleet (CINCPACFLT) S/N 0227 12 September 1967.

74 McKelvey Cleaver, Thomas. *The Tonkin Gulf Yacht Club.* Osprey Publishing 2001.

75 Roblin, Sebastien. *How America's Nuclear Skywarrior Saved Hundreds Over Vietnam.* The National Interest 3 March 2021.

76 Duthie memorialized this event in his book *Return to Saigon—A Memoir.* Another excellent book is Peter Fey's *Bloody Sixteen* that details CVW-16, which suffered the highest loss rate of any carrier air wing during the Vietnam conflict.

77 Morgan, Rick. *A-3 Skywarrior Units of the Vietnam War.* Osprey Publishing 2015.

---

[78] *Oriskany* would suffer several significant mechanical problems as a result of their hasty departure from the west coast. The most significant was the loss of two propellers and one shaft, necessitating extended stays in Yokosuka to make repairs.

[79] Navy A-3 losses by year: 1964-1 1965-2 1966-4 1967-6 1968-1 1969-2 1970-2 1971-1 1972-0 1973-1. These numbers represent both combat and operational losses. The Air Force lost a total of 17 B-66 Destroyers during the conflict, resulting in the deaths of 28 crewmen.

[80] There were typically a dozen LSOs within an air wing, two per squadron in addition to two from the air wing.

[81] The axle stub on the A-6 landing main gear was actually a mechanical fuse. On a hard landing the weld was designed to fail before the strut at the trunnion fitting in the wing box. This was to prevent the oleo from getting pushed thru the wing box, which has a fuel cell and thus a potential for fire or major, unrepairable damage to the aircraft.

[82] The immense B-52 Stratofortress was colloquially referred to as a "BUFF," an acronym for "Big Ugly Fat F****r".

[83] Clark was the grandson of fabled WWII Army General Mark Clark, whom Winston Churchill nicknamed "The American Eagle".

# Chapter VI

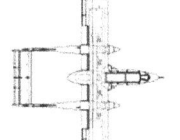

# Just the FACs
## The Long, Last Day

Numerous combat missions flown by Navy aircraft in the Vietnam theatre were under the direction of a Forward Air Controller, one of the true great air warriors of the conflict. Specialized FACs from all military branches flew missions twenty-four hours a day in North Vietnam, South Vietnam, Cambodia, and Laos. They often flew low over the jungled battlefields, locating and marking targets, coordinating close air support, spotting for artillery, assisting in the rescue of downed airmen, gathering intelligence, and other dangerous tasks. They worked directly with troops on the ground, and combat aircraft in the air, playing a crucial support role. Many a soldier, sailor, and airman who survived ground combat in Vietnam owe their existence to these indomitable men, 223 of whom lost their lives, often while trying to protect others.

Aerial FACs have existed in some form since the Civil War, when they rose above the battlefields in lighter-than air balloons. However, it was not until nearly a century later, during the Korean conflict, that their role became fully developed, and the critical part they played in close air support duly appreciated.

There were several notable aircraft that flew FAC missions in the Vietnam theatre. Most prominent were the piston-powered O-1 Birddog and O-2 Skymaster, also known as the "Oscar Deuce". Later in the conflict, the twin-turboprop OV-10 Bronco gained prominence as a FAC. There were also "Fast FAC" squadrons call sign "Misty" that flew the F-100 Super Sabre, and call sign "Stormy" that flew the F-4 Phantom. These specialized FACs primarily worked North Vietnam and parts of Laos, in areas considered too dangerous for piston or turboprop aircraft that were lightly armed. While the O-1 pilot seldom had anything more than a sidearm and rifle at his disposal, the O-2 and OV-10 featured mounted machine guns and other external munitions. When in a tight spot these daring, indomitable pilots were known to engage the enemy at close range, with whatever firepower they could muster.

Once a FAC located and marked a target with a "Willie Pete" rocket, a "fast mover" under their direction, commonly a turbine-powered fighter or attack aircraft such as the F-4 Phantom or A-4 Skyhawk, would come in fast and hard. These fast movers packed tons of high-explosive ordnance, which the FAC would direct, often with pinpoint precision and great effect. What aircraft were sent a FAC's way depended on the target. When fast movers checked in with the controlling authority, they stated their weapons load-out, which could range from guns and rockets to Mk general purpose bombs, Mk.77 incendiary bombs, and CBU cluster bombs. When a FAC requested air support they were assigned fast movers with appropriate munitions for their targets. For trucks, Mk.82s worked well, while CBU-24 cluster bombs with Mk.20 Rockeye bomblets worked best against tanks and armored vehicles.

FACs faced many monumental challenges in the theatre. One such challenge was that the enemy knew fast movers had to be under the control of a FAC to work effectively. The FAC was essentially their eyes, and if removed the fast movers could not do their job. Therefore, the FAC was commonly perceived as a threat, and the enemy would not hesitate to open fire on them.

Air Force 1st Lieutenant Adam West deployed to Vietnam as a FAC in April 1972, just as North Vietnam launched the Easter Offensive and crossed the DMZ into South Vietnam. Arriving to the 549th Special Operations Wing/20th Tactical Air Support Squadron at Da Nang flying an OV-10A call sign "Covey 115," West was initially assigned to an AO in central Laos. FAC pilots flying over Laos often lamented they felt like the "Laotian Highway Patrol," and had patches made for their flight suits that reflected that sentiment.

He flew missions in Laos until June, when he was brought back to South Vietnam due to a shortage of in-country FACs. West subsequently patrolled MR1, an AO south of the DMZ, north of the Quang Tri River and Route 9, west of Route 1, and east of the Laotian border. This included Con Thien, Dong Ha, Cam Lo, and Khe San, areas actively under NVA/VC attack during the offensive. Once airborne from Da Nang, West would check in with I Corps Direct Air Support Center (DASC), call sign "Big Look," alerting them to his presence. Big Look would instruct him to contact USAF Tactical Air Control Party (TACP) for South Vietnam, which was located in Hue, call sign "Trail Control". When West needed air support, Trail Control would send him assets, which could be aircraft from any military branch.

1st Lieutenant Adam West with OV-10 Bronco at Da Nang.

West flew dozens of missions in MR1 over the next five months, and in November he participated in one of the most consequential actions of the conflict—the Battle of An Loc. It was a night during during this pivotal battle that a panicked radio call came from a relatively new FAC that he had trained. On his first solo mission in South Vietnam, enemy forces were threatening to overrun an ARVN base, and the first-tour pilot was in over his head and badly needed assistance.

Breaking station from the AO he was working and heading north, West descended through an undercast layer a mile thick into a narrow valley, with soaring mountain peaks on both sides that were obscured by cloud cover lingering 1,500 feet above ground, making air support difficult, and dangerous. Within a few minutes of arriving, he had two F-4s ready to come in. However, with low cloud cover, obscured peaks, and AAA fire, the risk for them was great. He informed the two Phantoms of the difficulties, and left it to them whether they would make a run. Despite the challenging circumstances, the Phantoms agreed to give it a try. West remembers his instructions:

> I told then to come in tight, wingtip to wingtip, and also informed them there would be another OV-10 holding to the east, which was the new pilot. I instructed one aircraft to lay his ordnance at the wire of the base perimeter, and the other to lay his ordnance down on the hillside, where enemy guns were located. I cleared them in hot, and told them to be ready to drop their loads. They came in about a mile behind me and hit their targets. I ended up working six more aircraft that battle, which drove the NVA back and temporarily prevented the base from being overrun.

While working the battle, West noticed a blinking light in his cockpit, which turned out to be a low fuel warning indicator. Knowing he did not have enough gas to make it back to Da Nang, he diverted to the American air base at Pleiku, located in the central highlands. Another OV-10 FAC, piloted by Lieutenant Charlie Yates, joined up with him to make sure he landed safely, or call for rescue if he did not.

Landing at Pleiku was difficult, as night had fallen and the area was actively under attack. Adding to the challenges, beyond overrun areas of the runway were mine fields. West was doubtful he had enough fuel for a second landing attempt, so if he was not able to set it down on the first try he would likely have to eject over an area filled with enemy troops. He told Yates to "chop 'em and drop 'em," meaning get the airplanes stopped on the runway as quickly as possible. The two Broncos touched down and came to a halt at the very end of the overrun area, with mines not far beyond.

West and Yates refueled, however there was another problem. The runway at Pleiku was 6,000 feet, while a fully loaded Bronco could require up to 8,000 feet, leaving a 2,000 foot deficit. Fortunately, a procedure had been established for this very scenario. West parked a jeep at the 2,000 foot marker, and had the ground crew park a fuel truck at the end of the runway, where the overrun area began, providing visual cues for his takeoff. After running up the engines to full power, West released his brakes, and the Bronco began rolling. When he reached the jeep he retracted his landing gear, and let the nose raise only slightly, forcing the aircraft to stay in ground effect until enough speed was gained to climb. If the effort was not timed perfectly, the aircraft could stall or settle back on the ground beyond the runway, detonating the mines. In that case procedure dictated that once the aircraft came to a halt the pilot, if able, would egress from the cockpit, climb to the empennage, and walk back to the runway via the path of exploded mines. Fortunately, both West and Yates were able to safely takeoff from Pleiku, and returned to An Loc to continue providing FAC support to the besieged base. When they arrived, they discovered the low cloud cover had dissipated, which unleashed the devastating firepower of an Air Force AC-130 Spectre gunship that had appeared during their absence. Both men returned safely to Da Nang later that night.

On 18 December 1972 the United States launched *Operation Linebacker II*, also known as the "Christmas Bombings". The following day, West lost a close friend, fellow 20th TASS OV-10A FAC Captain Francis "Frank" Egan, call sign "Covey 64". In the backseat of Egan's Bronco that day sat Marine Corps Captain Jon Patterson, callsign "Wolfman 44," part of the Artillery-Naval Gunfire Liaison Company (ANGLICO) directing naval gunfire on numerous NVA elements advancing south across the DMZ.

Patterson had hastily arrived to SEA when NVN launched the Easter Offensive at the end of March. He and several other Marine ANGLICOs were attached to Air Force TASS squadrons at Da Nang, flying in the back seat of OV-10 FACs. ANGLICOs spotted for targets and coordinated gunfire from Navy boats in the gulf, among them in December 1972, the heavy cruiser USS *Newport News* (CA-148). As American combat units had left South Vietnam some years before, Jon was working with RVN Marines on the ground that had U.S. military liaisons embedded. Having been in theatre for both *Linebacker* and *Linebacker II*, the last two major air offensives of the conflict, Patterson had seen plenty of action. Before the fateful flight of 19 December, he had ejected from a Bronco crippled by ground fire not once, but twice. He recalls those first two ejections:

On the first ejection, we had taken a hit from an SA-7 in one of the engines. The pilot shut down the damaged engine, and everything seemed fine after that. We were very close to Da Nang when the pilot lowered the landing gear, and the airplane began to shake violently, with pieces of the airframe falling off.

The second ejection was much the same, we were hit in one of the engines by an SA-7. In addition to the damaged engine, we lost all hydraulics, resulting in the aircraft becoming uncontrollable. On both those ejections the pilot and I escaped serious injury, and were rescued.

On 19 December, Egan and Patterson departed from Da Nang on a daytime flight to Quang Tri Province. While flying south of Dong Ha, they were hit by an SA-7. Though neither man was injured by the missile, shrapnel inflicted enormous damage to the Bronco. Egan immediately turned east to get feet wet, where their chances of rescue were good.

Army Captain Warren Fuller was flying a U-21 that day, a military version of the Beechcraft King Air, call sign "Vanguard 969". Flying for the Army Security Agency, he was triangulating North Vietnamese radio signals to locate troop concentrations. While he had never met Egan in person, they were nevertheless friends, and as he puts it, "Sky Brothers":

> The first time I "met" Captain Frank Egan was on an early morning mission. I had just checked in with the AO, and asked to work at Angels 10 (10,000 feet MSL). Frank then chimed in and informed me that he was working at Angels 8.5 (8,500 feet MSL).

We wished each other luck, and that was that. About 30 minutes later, as we were working a target, I had this uneasy feeling to look to my left. To my surprise, there was an OV-10 just off of my wing. A conservative estimate would put our aircraft about two feet apart. My heart immediately lodge in my throat, and I could hardly breathe. He looked at me, and gave a thumbs-up as he veered off to his left. The ensuing "chatter" we had on the radio would make this story XXX rated. Suffice to say, we became instant friends.

Our missions were four hours in length, and we were given Air Force in-flight lunches. Frank would routinely ask me what I was having for lunch. I always tried to get the tuna fish lunch, which also came with a can of peaches, which I hated. Our standing joke was that I'd slide the peaches out to the end of my left wing for him to pick up at his leisure.

Fuller heard the distress call from Covey 64 and immediately took action:

> Frank had taken a hit from an SA-7 and was heading to the coast, so that he and Wolfman 44 [Patterson] could punch out. I immediately got a visual on the aircraft and started descending towards them, keeping the OV-10 in view at all times. They were losing altitude, and Frank told me that they would have to eject at 800 feet.

Burning (top) and submerged (bottom) OV-10 of Frank Egan and Jon Patterson just off the beach, south of the Cua Viet River 19 December 1972.

While following him, I declared myself as the on-scene commander of the rescue effort, and established radio contact with resources I thought could help. There was a Navy ship headed in our direction to lend support, a flight of Huey helicopters from Da Nang for pickup, a local ground commander who was in the vicinity of the beach, and a pair of jet fighters who were in our general area that may have been working with Frank earlier.

As Egan and Patterson approached the coast, they punched out at 800 feet, but I only witnessed one parachute deploy. Wolfman 44 contacted me when he got on the ground and told me that Frank's parachute never deployed, and he appeared to be dead. I was later to find out that a failure of a D-ring prevented his parachute from opening, and he was likely killed upon impact with the ground.

It was a busy, frantic day for rescue forces. Five other aircraft had been shot down—three B-52 BUFFs, an F-111, and a Navy A-7. SAR resources were stretched thin, and the area where Patterson was located, on the coast south of the DMZ, was very hot and therefore dangerous to any helicopter attempting an extraction.

While on the ground refueling their UH-1 Huey at Tan My Island, eight miles northeast of Hue, Army Captain Joe Bowen and his crew of six heard Fuller's mayday call. Bowen was an Air Mission Commander (AMC) with F Troop, 4th Cavalry, call sign "Centaur 3".

On alert for CSAR, they immediately lifted off for the thirty-five mile flight, radioing the remainder of their unit helicopters in Hue for assistance. Believing they were one of several helo assets that would be responding to the call, in reality none of their compatriots in Hue heard their transmission. Bowen and his crew were on their own.

Army Air Cavalry Captain Joe Bowen (standing) and others who rescued Jon Patterson and recovered the body of Frank Egan 19 December 1972.

He remembers that while en route, the ARVN liaison they had onboard was continuously yelling "You must turn around!" as they headed into a nest of NVA forces. Flying just 50-100 feet above the ground, when they reached Patterson they found him being protected by a group of Vietnamese soldiers. Though these soldiers were dressed in black and carried AK-47s, none of them fired a shot.

They loaded Egan's body onto the Huey, and in appreciation Bowen and his men gave them some "lurp" Long-Life Ration Packets (LLRP) and water. The helo then departed for the headquarters of the 1st ARVN at Hue, where Egan was officially declared deceased. To this day it is unknown if those mysterious, helpful soldiers in black were NVA who had deserted, or South Vietnam forces deceivingly dressed as the enemy. Regardless, without their help Patterson, and perhaps the crew of Centaur 3, would not have survived. The final tally of American air losses that December day were 8 KIA, 8 POWs and 8 rescued.

## The Long, Last Day

Adam West flew his 177th and final combat mission on Saturday 27 January 1973. Taking off from Da Nang, the load-out of his Bronco included four M60 7.62x51mm machine guns, and 28 Willie Pete rockets. With parties on the verge of a peace agreement in Paris, a ceasefire and cessation of hostilities was expected at any moment. However, past rumors of a peace agreement had turned out to be false, so for the 20th TASS, and other FAC units, it was business as usual.

As West lifted off from Da Nang, *Enterprise* was on Yankee Station, launching what would be the final Navy combat sorties of the conflict. Preparing for a mission that day was Lieutenant Commander Ernie Christensen, an F-4J Phantom pilot, and operations officer with VF-142. He had returned to combat duty two years prior, after two Vietnam tours with A-4 squadron VA-113 on *Enterprise*, and a 1969-1970 tour with the Blue Angels. He had been designated flight leader by squadron Skipper Commander Tom Bruyere that day, as Ernie had more combat time in theatre than anybody else in the squadron, including the skipper.

The men on *Enterprise*, particularly in the air wing, were also well aware of the progressing diplomatic efforts in Paris. During a pre-flight briefing with his three VA-142 wingmen and their RIOs, Christensen stressed the need to play it safe, and not take any unnecessary risks.

Also aboard *Enterprise* flying combat missions that day was Commander Harley Hall, who was XO of sister F-4 squadron VF-143. Before his combat assignment to the *Pukin Dogs*, Hall had been skipper of the Blue Angels, where he and Christensen served together in 1970. During that time they had become close friends and confidantes, with an immeasurable amount of respect and admiration for each other.

Hall and his RIO Lieutenant Commander Phillip Kientzler were near the fantail of the carrier, pre-flighting a Phantom (BuNo. 155768) for their second sortie of the day. Christensen, on the way to pre-flight a Phantom (BuNo. 155774) for what would be his 360th and final combat mission of the conflict, stopped by. He jested with Hall about neither of them "bagging" a MiG before the anticipated ceasefire. Christensen also recalls his parting words to his friend:

> I mentioned to him that wasn't it something, after all these missions, that we're both here flying the last sorties of the conflict before the anticipated ceasefire. More so, that we were both still "present and accounted for"! We laughed that we had survived, then manned up for the mission.

Christensen, along with his RIO Lieutenant Harry Hunter and three VF-142 wingmen, call sign "Dakota," launched from *Enterprise* ahead of Hall and his wingman. Once airborne, the flight of four turned to a heading that would take them to the north coast of South Vietnam.

They contacted Trail Control, who instructed the flight to proceed to an area northeast of Quang Tri, south of the DMZ, and rendezvous with a USAF OV-10 Bronco (SerialNo. 68-3806) call sign "Nail 89," piloted by 1st Lieutenant Mark Peterson and Captain George Morris, Jr., from the 23rd TASS/54th SOW based at Nakhon Phanom, Thailand. Although USAF OV-10 FACs typically carried a crew of one, on this mission Morris was riding along in the back seat as an extra set of eyes. Christensen remembers when he first came on frequency to check in with Peterson and Morris in Nail 89, their response echoed sentiments from his pre-flight briefing before launching from *Enterprise*. "Roger *Ghostrider*, just a point" they opined, "this could be the last one, don't do anything stupid!"

After he affirmatively acknowledged the warning, Dakota flight was directed by Nail 89 to the Battle of Cua Viet, a raging fight between NVA soldiers and *Task Force Tango*, a unit of ARVN Rangers. The battle was taking place east of Quang Tri, on the south bank of the mouth of the Cua Viet River. The Rangers had pushed north towards the DMZ, into territory the enemy held since March that year, and were encountering heavy resistance. In a firefight, they desperately needed close air support. Christensen discussed the target with Peterson and Morris in Nail 89 before making the decision to send his three wingmen to 12,000 feet, out of range of most ground fire, while he made the first run-in to the target to evaluate ferocity of the defenses, and whether other targets existed in the vicinity. After Nail 89 laid down a line of smoke to mark the target, Christensen rolled in and delivered three Mk.82 bombs, barely escaping AAA fire as well as an SA-7 SAM.

Commander Harley Hall

Lieutenant Commander
Ernie Christensen

Once airborne, Hall and Kientzler call sign "Taproom 113," along with wingman Lieutenant Terry Heath and RIO Lieutenant Phil Boughton call sign "Taproom 114," also contacted Trail Control in Hue. They were instructed to proceed to an area eight miles northwest of Quang Tri City, and rendezvous with a USAF OV-10 Bronco FAC (SerialNo. 15-5624) piloted by West, call sign Covey 115.

After making contact, the Phantoms were directed by West to a convoy of 15 North Vietnamese trucks near Thon Bai An, northwest of Dong Ha. West warned them of heavy AAA fire from 23mm, 37mm, and 57mm guns located south of the DMZ, and what appeared to be 85mm north of the DMZ, as well as numerous SA-7 missiles that had been fired at him. He also briefed Taproom of the bailout area, which was due east and feet wet, where SAR aircraft were orbiting. After marking the convoy with a Willie Pete, he cleared Hall and Kientzler hot to the target, instructing them to make their runs south to north, so any hung or tossed bombs would not be a threat to friendly forces located about twelve miles south.

Hall made his run, releasing six Mk.82 bombs, hitting one or two of the trucks. Heath then made his first run, releasing the same bomb load, with the same result. After their initial runs, both Hall and Heath reported heavy AAA while pulling out of the target. Cleared by West to make another pass, neither aircraft was able to score hits due to both broken cloud cover and having to constantly jink to avoid AAA, which was intensifying. West, orbiting at 9,500 feet, was also having to jink to avoid AAA, ducking in and out of the clouds to throw off the tracking of gunners. Though the belly of his Bronco had armor, the sides and windscreens lacked any such protection.

Commander Hall made a third pass, pickling another ripple of Mk.82s in a hailstorm of incessant and deadly ground fire. Pulling out of the target, the distinct sound of AAA hitting the fuselage was followed by the illumination of the master caution indicator in the cockpit. The controls of the jet began to turn sluggish, and fire appeared on the port wing. Taproom 113 called out mayday, announcing they had been hit on pull-out, and were headed feet wet for the coastline and the South China Sea, where SAR forces were located. Hall and Kientzler knew they had little time. While the F-4 was famous for its blazing speed, it was also infamous for becoming a brick when both engines were lost. The glide ratio was very poor, only being able to travel about 1.2 miles across the ground for every 1,000 feet of altitude lost.

Wingman Terry Heath, several miles southwest, asked for a position report or flare, so he could pinpoint their location. Frantically searching the sky, he spotted Taproom 113 to the northeast, fire blazing from the trailing edge of the port wing. He radioed Hall, letting him know he had a visual on them. They were on fire and needed to head to the coast and feet wet, where their chances of rescue were good. Adam West also broadcast a mayday and turned east towards the coastline, where Hall should be headed. He then contacted the USAF EC-130E ABCCC covering South Vietnam, call sign "King," to start coordinating a rescue effort.

Peterson and Morris in Nail 89 had laid down another line of smoke, and Christensen had just made a second run on the target, when Taproom 113 entered their AO near the Cua Viet, desperately trying to reach the relative safety of feet wet. Nail 89 made a radio call stating an F-4 had entered their area east of Dong Ha, fire blazing. West heard this and it surprised him, as he had assumed Hall would make a beeline for the coast, which was due east from where they were hit.

Instead, Hall had circled the target and was now southeast, heading for the water. Turning to intercept, West heard Hall announce on the radio they were ejecting. Christensen witnessed the two men punch out of the burning Phantom. It was a sight he would never forget:

> As I was delivering my last three Mk.82 bombs and avoiding another SA-7, on the guard frequency I heard a mayday of "Aircraft on fire—head for feet wet".
>
> I looked west, above and near, and saw an F-4 on fire. In seconds I saw an ejection seat fire, and parachute in the air a couple of miles away. The plane started to roll, yaw, and descend. Seconds later I saw another seat fire out of the aircraft.

The two men had barely made the coast when the Phantom began to roll uncontrollably, passing 60° of bank. They were at 4,000 feet when RIO Kientzler rocketed out of the crippled fighter, followed seconds later by Hall. West spotted the parachutes, orbiting the first one in his line of flight which was Kientzler, losing sight of Hall due to cloud cover.

The unmanned and aflame Phantom continued to roll, pitched down, and impacted the ground nearly vertically, resulting in a tremendous fireball.

West made radio contact with Kientzler (Taproom 113 Bravo) as he descended in his parachute, who stated he was okay physically, but was being fired at by ground forces. He believed Hall (Taproom 113 Alpha) had been shot, as he was limp in his parachute. "I think Alpha is dead!" Kientzler called out on the radio. West remembers:

> I rogered him back, saying I understood Alpha had been shot and was probably dead. Then I told him I thought I might be drawing the ground fire and was backing off.
>
> Kientzler said they had been receiving ground fire since ejecting, and it wasn't me they were shooting at, it was them. He then stated he too had been shot. That was the last communication I had with the crew of Taproom 113.

An on-shore breeze blew the parachutes west, back to landfall. Ducking in and out of clouds, and jinking to avoid AAA and SA-7s, West did not witness the men come down about a hundred feet apart, near an enemy-occupied village on a barren island where the Dam Cho Chua and Cua Viet Rivers meet, approximately five miles southwest of the coastline. This location, known as the "Fingers," was very close to the battle that Nail 89 and Christensen were supporting.

Wingman Heath in Taproom 114 had also called a mayday on the guard frequency when West shouted out another SA-7 missile headed towards them, passing just under the nose of their F-4. Disregarding the danger, Heath descended to 3,000 feet where they got a visual on the men in their parachutes. He then descended further, to a dangerously low 1,000 feet, well in range of ground fire. From there he saw the two men make landfall, witnessing Hall release and discard his parachute and begin running, making it likely he had been knocked unconscious during the ejection, or was only appearing as such to avoid being shot during his descent. Kientzler, who had been shot in the thigh during his descent, hung limp in his harness. West shouted out yet another SAM warning on the radio, with Heath breaking hard right, resulting in another near miss.

After witnessing the ejections, Christensen realized it was his friend Harley Hall. He visually tracked the two men as they descended, also witnessing Hall detach from his parachute and begin running. He then climbed to 5,000 feet and decided to send his three VF-142 wingmen, who were orbiting at 12,000 feet, back to *Enterprise* while he stayed on-scene to provide support. Though his Phantom had no guns and he was "Winchester" on ordnance, he was still determined to stay and help in any way he could, even if just facilitating communications.

ABCCC King began vectoring aircraft to the area to assist in the rescue effort. Since the men were down in the AO of Nail 89, Peterson and Morris, who had been orbiting with West south of the river at 9,500 feet, took the on-scene lead in the rescue. Nail 89 radioed West, stating they were descending lower to try and acquire a visual of the men on the ground. West did not believe that a good idea:

> I warned them not to. There was too much AAA in the area, and they had been firing a lot of SA-7s. Nail 89 acknowledged, but said they were going to go down for a look anyway. They started a run-in from the southwest, and had descended to about 4,000 feet.

That is when I saw an SA-7 fired at them from their six o'clock position. I called out "SAM, SAM, SAM! Break left, break left, break left!" They broke left, and popped a flare, which the missile appeared to bite on. Then, for some inexplicable reason, they broke back to the right, into the SA-7. It hit them in the right engine.

Christensen, by this time orbiting at 12,000 feet over the unfolding situation, also saw Nail 89 get hit, roughly ten minutes after Taproom 113 had gone down:

I heard Nail 89 tell West that they were going to go lower to see if they could get a visual on the men. They rolled nose-low and descended, and within five seconds and before they had reached their altitude for a search, an SA-7 streaked up from the ground and Nail 89 took a direct hit. The Bronco began tumbling, followed by several panicked radio calls from one of the men, I'm not sure who, stating "I can't get out...I can't get out!"

I would guess their altitude was 400 to 600 feet. After the radio calls, I saw two parachutes emerge from the OV-10. After the second parachute, the aircraft impacted the ground and caught fire.

Heath and Boughton in Taproom 114 also watched helplessly as the SA-7 impacted Nail 89, forcing the two men to eject. They observed about 30 enemy soldiers firing on them as they descended, and began converging on them after they landed near, but not on the same island where Hall and Kientzler appeared to make landfall. West was surprised by the two parachutes that emerged from Nail 89:

> It was not until I saw two good parachutes that I realized Nail 89 was carrying a crew of two. Since OV-10 FACs normally flew alone, I had presumed only one pilot. I put out another mayday call, then contacted King and Trail Control, informing them I had another aircraft down, and now four men were on the ground needing rescue. I told them to contact Da Nang, and get another FAC launched as my fuel state was starting to get low.

Da Nang initially denied West's request for a replacement FAC. West replied that the situation was critical, and another FAC needed to be launched as soon as possible. At this same time, a call came on the radio from King, stating they had lost two of their four engines, and were RTB. King then directed all aircraft in SEA on frequency to contact West in Covey 115, who already had his hands full with coordinating the rescue of four men on the ground amid heavy AAA and SA-7 SAM fire. If that wasn't enough Hillsboro, the ABCCC for Laos, also began vectoring aircraft to him.

As rescue forces of Corsair attack jets and Jolly Green rescue helicopters began checking in with West, Morris (Nail 89 Bravo) made contact with him on the radio, stating he was okay and had a visual on Peterson (Nail 89 Alpha), who also appeared okay. West replied, informing Morris that a rescue effort was being put together, with aircraft checking in. Morris radioed back, saying Peterson had been spotted and was about to be captured. The next gut-wrenching radio call from Morris was "Oh my God, they just killed him!" West rogered back, asking him to repeat his last transmission. Morris then stated "Alpha is dead! They just killed him!"

With ground fire still intense and uncertainty about the men, their conditions, and their locations, West could not bring rescue forces in yet. No matter how tempting, the risk of losing more men and aircraft was just too high. However, word was spreading rapidly amongst friendly forces in the area about the evolving situation. American military liaisons embedded with South Vietnamese ground forces radioed West, saying they were going to try and effect a rescue from the south. In addition, the Navy Corsairs and Air Force Jolly Greens who had earlier checked in with him were joined by additional combat aircraft, including a Navy A-6 Intruder and a dozen Phantoms—six Air Force, four Navy, and two Marine Corps. As the aircraft checked in, West furiously scrambled to keep track of them, as well as the situation on the ground, all while evading AAA and SAM ground fire. As each aircraft checked in, he noted their location, weapons load, and fuel state. Should an aircraft run low on fuel, he also had to arrange tankers, as well as routes to and from. He stacked them one on top of another in 500-foot intervals, and as more aircraft checked in, the task became monumental.

As he was scrambling to keep on top of the rapidly evolving situation and coordinate a rescue, Morris radioed again. West recalls the exchange:

> Morris stated, "They see me! They are coming for me! I'm going to surrender".
> I rogered him back and repeated I understood he was going to be captured.
> He rogered me back "Yes, I am going to be captured".
> I stated "Roger POW?"

He replied "Roger, POW!"
I repeated "Roger POW".
He then screamed "Oh no!" and I heard what sounded
like a machine gun open fire, and his radio went dead.
That was the last communication I had with Nail 89.

West had little time to react to the disturbing, final transmission from Morris. In addition to the numerous aircraft that had checked in with him, American military liaisons on the ground radioed, stating they were encountering stiff resistance in their march north, and needed close air support immediately. However, with dusk approaching, an overcast cloud cover had formed at 2,000 feet above ground, making close air support dangerously difficult. With decaying visibility, impending darkness, and his fuel state reaching critical, West turned over his duties to a replacement FAC that had been diverted from MR-2, crossing paths as he was exiting and the replacement was arriving. Fortunately, he had trained the replacement, and he knew the area well. West fully briefed him on the situation on the ground, and aircraft that had checked in, stacked between 11,500 and 20,000 feet.

With his fuel tanks on empty, West requested a Jolly Green rescue helicopter that had arrived on-scene to escort him back to Da Nang, in case he ran out of fuel en route. After landing safely, he didn't need to shut down the engines—they quit on their own as a result of fuel starvation. When the aircraft was refueled, the crew chief informed him that the Bronco took more fuel than it was supposed to hold.

The heavy ground fire, cloud cover, darkness, and lack of further communication from the men, precluded any further rescue attempts. Shortly, word of the peace agreement and ceasefire spread, resulting in American aircraft once again all but disappearing from the skies of

North Vietnam. Thus, Hall and Kientzler were the last Navy MIA and POW of the conflict, respectively. Peterson and Morris carry the sad distinction of being the final two USAF KIA before the peace accords were signed. Shortly after the events that day, Heath and Christensen were briefed by intelligence officers onboard *Enterprise*. Peterson and Morris had been found by friendly forces, murdered. Their remains were not recovered.

After Kientzler was captured, he was told by a North Vietnamese soldier that Hall had been shot and killed, despite subsequent intelligence indicating that he had been captured alive. Kientzler was released several months later during *Operation Homecoming*. In 1993 the mortal remains of Commander Hall, consisting of three teeth, were repatriated. They were positively identified in 1994.

After returning to Da Nang, West was debriefed, giving a long and detailed account of the events, correctly assessing that three men had lost their lives. While he had witnessed many emotionally painful events during his deployment, more than fifty years later the panicked voices of Taproom and Nail continue to reverberate in his mind, their cries never to be forgotten.

# Contributors

**Chapter I Ray Gun 502 is Burning**
Captain Eugene "Red" McDaniel, USN (ret.)
Lieutenant Commander Nicholas "Nick" Carpenter, USN
Dave "Dobie" Cable, USN
Dr. William "Mitzi" Gaynor DVM, USN
Ken Van Lue, USN
Edward "Fast Eddie" Leonard, USN
Edward "Sadistic" Sadowski, USN
Pete Carrothers, USN
Captain Paul Daley, USNR (ret.)
Colonel Jim Craig, USAF (ret.)
Lee Cargill United States Naval Academy Class of 1963
Jim Ring United States Naval Academy Class of 1963
Kent Maxfield United States Naval Academy Class of 1963
Bill Schultz Grumman Aircraft Corporation (ret.)
Jade Nguyen Lyon Air Museum
Meredith Carpenter Page
Bob LaFramboise, USAF
Commander Bill Lindsey, USNR (ret.)

**Chapter II The Legend of Lucky 7**
Captain Philip Ryan, USN (ret.)
Pete Carrothers, USN
Captain George Clark USN, (ret.)
Commander Scott Reuther, USN (ret.)
Tom Joyner, USN
Jim Owen, USN
Bill Schaefer, USN
John Henry Fowlkes, USN
John "Bitz" Bitzberger, USN
Bill "Moose" Feldhaus, USN

Greg "Snoopy" Davison, USN
John Sutor, USN
Kathleen Ryan Sise
Cleta Humphrey
Commander Robert "Boom" Powell, USN (ret.)
Jay Pirotte

## Chapter III Battle Cry of the Stingers
Vice Admiral William "Bill" Bowes, USN (ret.)
Rear Admiral Ernest "Ernie" Christensen, USN (ret.)
Rear Admiral Jeremy "Bear" Taylor, USN (ret.)
Jay Greene, USN
Captain Tom "One Shot" Scott, USN (ret.)
Captain Tom "Boomer" Brown, USN (ret.)
Paul Adams, USN
Arne "Bud" Johnson, USN
Harry "The Rat" Welch, USN
Maureen Ellis
Stephen Gray, USN
Captain Dave Dollarhide, USN (ret.)
*The Skyhawk Association*
Debbie and Ray Loehner

## Chapter IV It's Showtime/The Continuing Saga of Ensign Thornton
Commander Terry "Jonas" Born, USN (ret.)
Jim Ritchie, USN
Doug Kindseth, USN
Commander Gary Thornton, USN (ret.)
Lieutenant Commander Jim Hollarn, USN (ret.)
Captain Paul Daley, USNR (ret.)
Jim Stillinger, USN
Captain Dave Hoffman USN (ret.)
Betty Ray Wilson
Captain Lewis "Scurvy Irv" Williams, USN (ret.)

## Chapter V Tales of the Whale
Commander Kenneth "Mike" Murphy, USN (ret.)
Commander Jon "Buz" Vaughters, USN (ret.)
Captain Gregg Bambo, USN (ret)
Captain Ed Gibson, USN (ret.)
Captain David Mason, USN (ret.)
Captain Terris "T-Bone" Hanson, USN (ret.)
Commander Howard "Nick" Nickerson, USN (ret.)
Grant "Scottie" Meiner, USN
Scott "Scottie" Atkins, USN
Captain Charles"Moth" Hanson, USN (ret.)
Captain Steve Kuhar, USNR (ret.)
Alec Schmidt, USN
*A-3 Skywarrior Association*

## Chapter VI Just the FACs/The Long, Last Day
Rear Admiral Ernest "Ernie" Christensen, USN (ret.)
Captain Adam West, USAF (ret.)
Jon Patterson, USMC
Warren Fuller, United States Army Security Agency (ASA)
Joe Bowen, United States Army
*Forward Air Controllers Association*
*National Army Security Agency Association*

# Photo Credits

**Chapter I Ray Gun 502 is Burning**

| | |
|---|---|
| Page 3 Kelly and Luck in Hoi An (top) | Luck Patterson |
| Page 3 Kelly and Luck in Hoi An (bottom) | Luck Patterson |
| Page 8 Ray Gun 502 at NAS Alameda | Luck Patterson |
| Page 14 Kelly 1963 USNA Graduation | Luck Patterson |
| Page 14 Kelly receiving his wings | Luck Patterson |
| Page 17 Intruder Ball | Luck Patterson |
| Page 20 Kelly atop A-6 | Luck Patterson |
| Page 20 Kelly and Red on Yankee Station | Luck Patterson |
| Page 21 Squadron Photo 1966 | Luck Patterson |
| Page 27 Christmas in ready room | Luck Patterson |
| Page 27 Happy times below deck | Luck Patterson |
| Page 30 Kelly aboard *Enterprise* in Pearl Harbor | Luck Patterson |
| Page 30 Slaasted, Johnson, Cable, and McDaniel | Luck Patterson |
| Page 34 Kelly, Carpenter, Slaasted and McDaniel | Luck Patterson |
| Page 34 Gaynor and Leonard | Luck Patterson |
| Page 35 Johnson and Cable | Luck Patterson |
| Page 35 Mallek and Barie | Jim Hollarn |
| Page 38 A-6 readying to launch | Luck Patterson |
| Page 38 Over Tonkin Gulf | Luck Patterson |
| Page 39 Cable and Johnson over Tonkin Gulf | Luck Patterson |
| Page 39 Kollman and Ky | Luck Patterson |
| Page 45 Ray Gun 502 on fire | Paul Daley |
| Page 45 Red and Kelly parachutes | Paul Daley |
| Page 54 Commander Barie | *Enterprise* 1966-1967 Cruise Book |

ACROSS THE WING

## Chapter III Battle Cry of the Stingers

|---|---|
| Page 150 Squadron photo | *Kitty Hawk* 1965-1966 Cruise Book |
| Page 155 Heinemann's hotrod | Jay Greene |
| Page 155 Mk.82 bombs away | Jay Greene |
| Page 151 Commander Dibble | *Kitty Hawk* 1965-1966 Cruise Book |
| Page 151 Commander Abbott | *Kitty Hawk* 1965-1966 Cruise Book |
| Page 151 Bill Ellis | Ellis family |
| Page 151 Commander Burnett | *Enterprise* 1966-1967 Cruise Book |
| Page 161 Jay Greene | Jay Greene |
| Page 161 Bear Taylor | Bear Taylor |
| Page 163 Squadron photo | Jay Greene |
| Page 166 Christmas in ready room | Bill Bowes |
| Page 170 AAA damage | Jay Greene |
| Page 170 AAA over Haiphong | Bear Taylor |
| Page 173 Dragon's Jaw | Jay Greene |
| Page 183 VPAF airfield (top) | Bear Taylor |
| Page 183 VPAF airfield (bottom) | Bear Taylor |
| Page 189 In formation | Jay Greene |
| Page 193 Firing a Bullpup | Jay Greene |
| Page 197 Bear Taylor, DePrez and Scott | Bear Taylor |
| Page 198 Vaih Hoc | Jay Greene |
| Page 200 SAM detonation | Bear Taylor |

vi

Page 202 Readying to launch       Jay Greene

Page 208 Pete Paine       Jay Greene

### Chapter IV It's Showtime/The Continuing Saga of Ensign Thornton

Page 231 Wilson and Ritchie       Doug Kindseth

Page 234 Commander Norman       *Enterprise* 1965-1966 Cruise Book

Page 234 Commander Schwartz       *Enterprise* 1966-1967 Cruise Book

Page 238 CAP over Tonkin Gulf       Jim Stillinger

Page 238 Over Mekong Delta       Doug Kindseth

Page 247 Squadron photo       *Enterprise* 1966-1967 Cruise Book

Page 251 Christmas 1966 (top)       Jim Stillinger

Page 251 Christmas 1966 (bottom)       Doug Kindseth

Page 253 Jim Hollarn       Jim Hollarn

Page 258 Firing rockets       Jim Stillinger

Page 261 AAA damage to windscreen       Terry Born

Page 265 Flying CAP       Jim Stillinger

Page 265 Art Cisson       Jim Stillinger

Page 273 Ready to launch       Jim Stillinger

Page 275 Doug Kindseth       Doug Kindseth

Page 276 Captain Holloway       *Enterprise* 1966-1967 Cruise Book

Page 276 Commander Rich       *Enterprise* 1966-1967 Cruise Book

# Glossary

| | |
|---|---|
| 1MC | Ship-wide public address system. |
| AA | Anti-Aircraft. |
| AAA | Anti-Aircraft Artillery. |
| ABCCC | Airborne Command and Control Center |
| AFB | Air Force Base. |
| AGL | Above Ground Level. |
| AGM 12B | Bullpup missile with 250lb. warhead. |
| AGM 12C/C2 | "Big" Bullpup missile with 2,000lb. warhead. |
| AGM 45.A1 | Shrike Anti-radiation missile with 150lb. warhead used against Fire Can and Fan Song radar. |
| AGM 45.A3 | Shrike Anti-radiation missile with 150lb. warhead used against Bar Lock and Fan Song radar. |
| AGM-62 | Walleye glide bomb with 250lb. warhead. |
| AIM | Air Intercept (air-to-air) Missile. |
| AIM-7 | Sparrow radar-guided AIM. |
| AIM-9 | Sidewinder heat-guided AIM. |
| AIO | Air Intelligence Officer. |
| *Air America* | A clandestine air force of the CIA. |
| Alpha strike | An attack to deliver large amounts of ordnance on a high-value target deemed of great importance using numerous aircraft types from multiple squadrons, each with specific roles. |

| | |
|---|---|
| AO | Area of Operation. |
| AOCS | Aviation Officer Candidate School. |
| AOM | All-Officers Meeting. |
| ARRS (USAF) | Aerospace Rescue and Recovery Service. |
| ARVN | Army of the Republic of [South] Vietnam. |
| ATC | Air Traffic Control. |
| Atoll | A Soviet reverse-engineered Sidewinder. |
| BARCAP | Barrier Combat Air Patrol. |
| BDA | Bomb Damage Assessment. |
| B/N | Bombardier/Navigator. The right-seater of an A-6 Intruder or A-3 Skywarrior, responsible for navigation, radar, and weapons. Typically not a pilot. |
| Bogie | Unidentified aircraft. |
| Bolter | When an aircraft landing on a carrier misses all four arresting cables. After applying full power the aircraft must become airborne again for another attempt at landing. |
| BOQ | Bachelor's Officers Quarters. Military lodging/billet facility for unmarried officers or officers on temporary duty. |
| Bullpup | See AGM-12. |
| BuNo. | Bureau/Build Number. The serial number of a USMC/USN aircraft. |
| BVR | Beyond Visual Range. |
| CAG | Commander Air Group. |
| CAP | Combat Air Patrol. |
| CAS | Close Air Support. |

| | |
|---|---|
| CBU-2/24 | Cluster Bomb Unit. |
| CG | Center of Gravity. |
| CINCPAC | Commander in Chief Pacific Fleet. |
| CMOH | Congressional Medal of Honor. |
| C/N | Crewman/Navigator. |
| CNO | Chief of Naval Operations. |
| CO | Commanding Officer/Skipper. |
| COD | Carrier Onboard Delivery. |
| Condition Watch | An aircraft fueled, armed, and manned to launch within a specific time period. "Condition 5" meant the aircraft could launch on five minutes notice. |
| Crown/Red Crown | Callsign for Navy ship in the Tonkin Gulf that provided radar coverage of the skies over the Tonkin Gulf and North Vietnam. |
| CRT | Cathode-Ray Tube. |
| CSAR | Combat Search and Rescue. |
| CTF | Carrier Task Force. CTF-77 operated in the Vietnam theatre from April 1964-October 1973. |
| CVW | Carrier Air Wing. |
| DEW | Distant Early Warning. |
| DFC | Distinguished Flying Cross. |
| Dixie Station | Location off the east coast of South Vietnam where Navy aircraft carriers launched missions against targets primarily in the Republic of [South] Vietnam. |
| DMZ | Demilitarized Zone. |
| DoD | Department of Defense. |

| | |
|---|---|
| DRV | Democratic Republic of [North] Vietnam. |
| ECM | Electronic Counter-Measures. |
| ELINT | Electronic Intelligence. |
| EMCON | Emissions Control. |
| EMP | Electromagnetic Pulse. |
| Empennage | The tail section of an aircraft with vertical and horizontal stabilizers, rudder, and elevator. |
| ET | Electronics Technician. |
| EW | Electronic Warfare. |
| Executive Officer (XO) | Second in command of a military unit. |
| FAC | Forward Air Controller. |
| Fan Song (SNR-75) | Radar used for SA-2 SAM guidance. |
| FASU | Fleet Air Support Unit. |
| Feet Dry | Over land. |
| Feet Wet | Over water. |
| FFAR | Folding Fin Air Rocket. See LAU 3/LAU 10. |
| Fire Can (SON-9) | Radar for 57mm and 100 mm AAA guns. |
| Flag Officer | Admiral or general grades O-7 to O-10. |
| Flap Wheel (SON-50) | Radar for 57mm anti-aircraft guns. |
| Fly-by-Wire | A flight control system that uses electrical signals to manipulate powered actuators on the control surfaces of the airplane. |
| FOD | Foreign Objects & Debris. |

| | |
|---|---|
| Ground Effect | When an aircraft flies within a wingspan distance of the ground, air deflected downward from the wings collide with the surface and rise back up, providing a cushion of air between the aircraft and ground which results in additional lift and improved performance. |
| Guard Channel | The guard frequency of 243 MHz was reserved for priority messages. It was also the frequency used by portable radios carried by airmen for rescue. |
| HUD | Heads-Up Display. |
| ICBM | Intercontinental Ballistic Missile. |
| ICS | Intercom System. |
| IFR | Instrument Flight Rules. |
| IMC | Instrument Meteorological Conditions. |
| INS | Inertial Navigation System. A navigation system where geographic location is determined using inertial sensors onboard the aircraft. |
| IOIC | Integrated Operations Intelligence Center. |
| IP | Initial Point. |
| Iron Hand | Specialized Navy aircraft such as the A-4 or A-6 that utilized the AGM-45 Shrike anti-radiation missile. |
| Iron Triangle | The most heavily defended area of North Vietnam. The northern points of the triangle were Hanoi and Haiphong, with the southern point being Nam Dinh. |
| JATO *(jay-tow)* | Jet-Assisted Takeoff. |

| | |
|---|---|
| JBD | Jet Blast Deflector. A hydraulically operated shield that raises behind a fixed-wing aircraft prior to launch, protecting other aircraft, personnel, and equipment from jet blast or prop wash. |
| JCS | Joint Chiefs of Staff. |
| JEST | Jungle Environment Survival Training. |
| Jink | An aggressive maneuver to throw off the optical aiming or radar tracking of AAA/SAMs. |
| Junior Officer (JO) | Officer grades O-1 to O-4. Within the Navy and Coast Guard, this corresponds to ensign, lieutenant (junior) grade, lieutenant, and lieutenant commander. Within the Air Force, Army, and Marine Corps this corresponds to 2nd lieutenant, 1st lieutenant, captain, and major. |
| K-13 | See Atoll. |
| KIA | Killed in Action. |
| LAU 3 | 2.75" Folding-Fin Aircraft Rocket (FFAR). |
| LAU 10 | 5" Folding-Fin Aircraft Rocket (FFAR). |
| LSO | Landing Signal Officer. |
| MACV | Military Assistance Command Vietnam. |
| MCAS | Marine Corps Air Station. |
| MER | Multiple Ejection Rack. |
| MIA | Missing in Action. |
| MiGCAP | MiG Combat Air Patrol. |
| Milk Run | An "easy" mission where little to no enemy action is expected. |

| | |
|---|---|
| M.117 | 750lb. general-purpose bomb (WWII/ Korea era). |
| Mk.24/36/50/52 | Maritime mines. |
| Mk.77 | 750lb. incendiary bomb. |
| Mk.81 | 250lb. general-purpose bomb. |
| Mk.82 | 500lb. general-purpose bomb. |
| Mk.83 | 1,000lb. general-purpose bomb. |
| Mk.84 | 2,000lb. general-purpose bomb. |
| MR | Military Region. |
| MTOW | Maximum Takeoff Weight. |
| NAS | Naval Air Station. |
| NATO | North Atlantic Treaty Organization. |
| NAVCAD | A USN/USMC program for those with at least two years of college or equivalent knowledge/experience. Initial training was at NAS Pensacola. Graduates of the program were commissioned a USN ensign or a USMC 2nd lieutenant and continued training to become naval aviators. Those who did not graduate were inducted as enlisted personnel. |
| NFO | Naval Flight Officer. Typically not a pilot. |
| NFWS | Naval Fighter Weapons School (Top Gun). |
| NLF | National Liberation Front. |
| NTPS | Navy Test Pilot School. |
| Nugget | A first-tour naval aviator. |
| NVA | North Vietnamese Army, same as PAVN. |
| NVN | North Vietnam/North Vietnamese. |
| O-Club | Officer's Club. |
| OinC (*oink*) | Officer-in-Charge. |

| | |
|---|---|
| OOD | Officer of the Deck. |
| *Operation Barrel Roll* | Operation targeting sections of the Ho Chi Minh Trail in northern and central Laos, near the Barthelemy Pass and Samneua. Handled primarily by the USAF. |
| *Operation Homecoming* | February 1973-April 1973 repatriation of 591 imprisoned American personnel from North Vietnam, South Vietnam, Laos, and China. |
| *Operation Steel Tiger* | Operation targeting sections of the Ho Chi Minh Trail in southern Laos including the Mu Gia and Ban Karai passes. |
| OPREP | Operational Report. |
| ORE | Operational Readiness Exercises. |
| ORI | Operational Readiness Inspections. |
| PAVN | People's Army of [North] Vietnam. |
| PECM (*peck-um*) | Passive Electronic Counter Measures. |
| Pipper | Projected Impact Point. |
| PIRAZ | Positive Identification Radar Advisory Zone. |
| POL | Petroleum, Oil and Lubricants. |
| POW | Prisoner of War. |
| PPDI | Pilot's Projected Display Indicator is similar to a HUD. |
| PSO (*pay-so*) | Pilot Systems Operator. Similar to a F-4 WSO with one major difference—the |
| PSO | was a fully qualified pilot. |
| Punch Out | Eject from an aircraft. |

RAG                    Replacement Air Group. A Navy or
                       Marine Corps aerial training squadron
                       from which replacement aircrew were
                       drawn.
RAN                    Reconnaissance Attack Navigator. The
                       back-seater of an RA-5C Vigilante,
                       responsible for navigation, electronic
                       warfare, and intelligence imagery.
                       Typically not a pilot.
RAT                    Ram Air Turbine.
Ready Room             A compartment on an aircraft carrier
                       where squadron members congregate,
                       receive briefings, and prepare for missions.
                       Normally each squadron has its own
                       ready room however they are sometimes
                       shared between squadrons.
Recce (*rek-e*)        Reconnaissance mission.
RESCAP                 Rescue Combat Air Patrol.
RHAW (*raw*)           Radar Homing and Warning gear.
                       Typically the AN/APR series.
RIO (*ree-oh*)         Radar Intercept Officer. A Navy or
                       Marine Corps flight officer, seated behind
                       the pilot in the F-4 Phantom, primarily
                       responsible for navigation, monitoring
                       radar instruments, radios, and other
                       aircraft systems. Typically not a pilot.
ROE                    Rules of Engagement. Criteria of
                       conditions set by command elements for
                       conducting military operations against
                       enemy forces.

| | |
|---|---|
| RTB | Return to Base. |
| RVAH | Navy Heavy Reconnaissance Squadron. |
| RVN | Republic of [South] Vietnam. |
| SA-2/S-75 | Soviet *Dvina* SAM air defense system. |
| SAM | Surface-to-Air Missile. |
| SAR | Search and Rescue. |
| SEA | Southeast Asia. See Vietnam theatre. |
| Senior Officer | Officer grades O-5 & O-6. Within the Navy and Coast Guard, this corresponds to commander and captain. Within the Air Force, Army, and Marine Corps this corresponds to lieutenant colonel and colonel. |
| SERE | Survive, Evade, Resist, Escape. Similar to JEST. |
| Sidewinder | See AIM-9. |
| SLR/SLAR | Side-Looking (Airborne) Radar. |
| Snake Eye | Bombs fitted with high-drag tail fins for retarded fall. |
| SOP | Standard Operating Procedure. |
| SOS | Special Operations Squadron. |
| Sparrow | See AIM-7. |
| Spoon Rest (P-12) | Radar used for SA-2 target acquisition. |
| Stall | When lift is no longer being produced by the wing/airfoil, caused by exceeding the Critical Angle of Attack (AoA). |
| SVN | South Vietnam/South Vietnamese. |
| TARCAP | Target Combat Air Patrol. |
| TASS | Tactical Air Support Squadron. |
| TDY | Temporary Duty. |

| | |
|---|---|
| TEWS | Tactical Electronic Warfare Squadron. |
| TFS | Tactical Fighter Squadron. |
| TFW | Tactical Fighter Wing. |
| Top Gun | See NFWS. |
| TPP | Thermal Power Plant. |
| TRS | Tactical Reconnaissance Squadron. |
| TRW | Tactical Reconnaissance Wing. |
| USAF | United States Air Force. |
| USMC | United States Marine Corps. |
| USN | United States Navy. |
| USNA | United States Naval Academy Annapolis. |
| USNTPS | United States Navy Test Pilot School. |
| VA | Navy Attack Squadron. |
| VAH | Navy Heavy Attack Squadron. |
| VC | Viet Cong. |
| VF | Navy Fighter Squadron. |
| VFR | Visual Flight Rules. |
| Vietnam Theatre | Republic of [South] Vietnam, Democratic Republic of [North] Vietnam, Kingdom of Laos, and Kingdom of Cambodia. |
| VNAF | Republic of [South] Vietnam Air Force. |
| VNN | Republic of [South] Vietnam Navy. |
| VPAF | [North] Vietnamese People's Air Force. |
| VPN | [North] Vietnamese People's Navy. |
| Walleye | See AGM-62. |
| WESTPAC | Western Pacific Cruise. The term used for naval deployment cruises to the western Pacific Ocean/SEA region. Typically 6-7 months in duration. |
| Willie Pete | White Phosphorus. |

| | |
|---|---|
| Winchester | Out of Ordnance. |
| WSO (*wiz-oh*) | Weapons Systems Officer. A USAF flight officer, seated behind the pilot in the F-4 Phantom, primarily responsible for navigation, monitoring radar instruments, radios, and other aircraft systems. Typically not a pilot. |
| Yankee Station | Location off the east coast of North Vietnam where Navy aircraft carriers launched missions against targets primarily in the Democratic Republic of [North] Vietnam and Laos. |

www.ingramcontent.com/pod-product-compliance
Lightning Source LLC
Chambersburg PA
CBHW051606120626
46551CB00014B/1690